H

# THE FUTURE WITH MICROELECTRONICS

# THE FUTURE WITH MICROELECTRONICS

**Forecasting the Effects of Information Technology**

## Iann Barron

INMOS

## Ray Curnow

**with other members of the Science Policy
Research Unit at Sussex University**

**THE OPEN UNIVERSITY PRESS
MILTON KEYNES**

The Open University Press
12 Cofferidge Close, Stony Stratford
Milton Keynes MK11 1BY, England

First published in Great Britain in 1979
by Frances Pinter (Publishers) Ltd.,
5 Dryden St., London WC2 9NW

First published in this edition by the
Open University Press 1979

ISBN 0 335 00268 4

**British Library Cataloguing in Publication Data**

Barron, Iann
    The future with microelectronics.
    1. Computers and civilization
    2. Microcomputers – Social aspects
    I. Title    II. Curnow, Ray
    301.24′3        QD76.9.C66

    ISBN 0–335–00268–4

Typeset by Texet, Leighton Buzzard, Bedfordshire and
Printed in Great Britain by Billing & Sons Limited,
Guildford, London and Worcester

# TABLE OF CONTENTS

*Foreword by Sir Ieuan Maddock*

Preface to the Paperback Edition
Preface

The importance of computing; Information technology;
The theory of information; Information as a resource
Communications systems; The office; The factory;
The home
The industrial revolution; Kondratiev long waves;
The information revolution

# FOREWORD

**Future of Information Technology**

In 1970 a group of prominent people was assembled by the O.E.C.D. Industry group to make a visit to Japan to examine and discuss what was then described as the 'Japanese economic Miracle'. I was one of this group and one of the most profound impressions I carried away was the clear view those responsible for policy had of the future of Japan. They talked in terms of Japan being the world's leader in the 'information based society' — a salutory signal that they at least had a clear view of where the main thrust of technology was going to be and their own place within it. All that has happened since has confirmed the clarity and relevance of their vision.

It is this vision that all of us must share. For as far as any outlook on the future of technology can reach — well into the next century, the ability to gather, record, organise, analyse and act upon information is going to be a dominant factor. What steam, steel, electricity was to the nineteenth century, information management and exploitation will be for the next half century if not very much longer. Not only is it the new 'raw material' of technology, it will inevitably become an essential ingredient of the fabric of the human society.

It is timely therefore to take stock and look to the future, not only in terms of the inner characteristics of the science and the technology of information management itself, but in terms of its impact on employment, education, social management and even political structure. The authors of this book are singularly well placed to undertake this task.

Sir Ieuan Maddock                                        February 1979

# PREFACE TO THE PAPERBACK EDITION

Since the study reported in this book was completed in January 1978, much has happened. In the U.K., a BBC Horizon programme, 'Now the Chips are Down' was broadcast early that year and referred to a 'deafening silence' on this subject. However, the Advisory Council on the Application of Research and Development (A.C.A.R.D.) was already at work, and produced its report on the application of microelectronics in September 1978. Amongst several other papers, the long awaited Think Tank (Central Policy Review Staff) report appeared in December 1978, although by then the Department of Industry had launched its Microelectronics Industry Support Programme (MISP) and Microelectronics Application Programme (MAP). The subject, too, had been discussed at all three political party conferences in the summer of 1978, and also at the TUC Conference of that year.

Abroad, 1978 saw the publication of the official French report by Nora and Minc called '*l'Informatisation de la Societe*', and various reports produced by trade union or other study groups in Europe are circulating. In Germany, the microprocessor was nicknamed 'the job-killer', whilst in Norway the formal establishment of union 'data-stewards' to put forward alternative plans for information-technology implementation was legally recognised.

December 1978 saw the setting up of a formal TUC study group on Technology and Employment, while AUEW-TASS and APEX have both produced substantial reports on the implication of computerisation. On the industrial front, the setting-up of INMOS, with

Iann Barron as strategic director, and supported by the National Enterprise Board, but accompanied by controversy over the use of public money in support of private enterprise, was followed by the joint GEC-Fairchild venture to produce microelectronics in the U.K. Other notable events have been the declared intention of Plessey to concentrate on office automation via its own private telephone-switching exchange development, and an apparent agreement between the British Post Office and the Post Office Engineering Union that no redundancies will ever follow future technical change.

Views on whether the advent of·information technology will actually produce very sharp dislocatory effects still differ, although it is my contention that the evidence to that effect is already present and growing. To a large extent, this debate is now being overshadowed in the U.K. at least by the now accepted recognition of continued industrial decline, set against a background of unstable and increasing energy prices triggered by the social upheaval in Iran. I would contend that all these issues are part of the same problem — that of economic, social and political adjustment proving increasingly difficult in a world of accelerating technical change potential. Perhaps the U.K. will be only the first of the industrialised countries to have to consider its future more systematically — perhaps I am wrong in believing that the relative swing to more conservative administration in many Western countries is a false search for a return to older, more stable conditions.

Be that as it may, there is now little doubt that information technology could be used to build a world in which automation could remove much drudgery, and release human creative potential for individually chosen better lives. There is equally little doubt that the problems of transition will almost certainly be severe, not least in the U.K. Viewed as the latest but certainly most powerful of a long series of technological developments, it might be as well to remember the words of Aeschylus: 'For it behoves a man to take account of his position at noon.'

I welcome this edition of *The Future with Microelectronics* by the Open University Press as a contribution to the debate as to where our society is going, can go, in the belief that the U.K., because it will be the first country to face this challenge, may help point the way to a peaceful transition to a better world.

Ray Curnow

# PREFACE

This study was carried out at the request of the Computers, Systems and Electronics Requirements Board (C.S.E.R.B.) of the Department of Industry in the United Kingdom. The Research Requirements Boards, of which C.S.E.R.B. is one, were set up following the Rothschild recommendations in 1971 of formalising a customer-contractor relationship into U.K. Government research and development spending, and act as proxy customers for this purpose in relation to governmental laboratories. The Boards include lay members from industry and commerce as well as civil servants with appropriate departmental responsibilities.

The research began in August 1976, with final amendments made in January 1978. Originally conceived as an eighteen-man-month study, to be completed in nine months, the work was carried out by the Science Policy Research Unit (S.P.R.U.) of the University of Sussex, with Iann Barron, Visiting Professor at Westfield College, London, leading. Ray Curnow led the work at S.P.R.U., which drew on overlapping and previous projects carried out there. The brief of the study was to carry out a technological forecast in the computing field with a suitable timespan, in order to provide one reasonably comprehensive and consistent view as to how that field might emerge. Most of the information put forward was based on the personal experience and reading of the two authors supplemented by private discussions, and no commercially sensitive or secure information was used, nor are references given in this early version. One or two of the technical developments foreseen are already well in train, others will probably not now take place — such is the lot of forecasters.

Ray Curnow                    Iann Barron                                   April 1979
S.P.R.U.                      INMOS

# CHAPTER ONE

# INTRODUCTORY SUMMARY

The question of future developments in information technology and their impact on the supply industry and the overall economy is of vital importance to everyone. This study is directed towards the technological implications of these developments and their consequences for policy.

## INFORMATION TECHNOLOGY
Contemporary information technology embodies a convergence of interest between electronics, computing and communications, which is being promoted by the rapid development of microelectronics. In the past, information has been handled by a variety of techniques, mainly paper based. The advances in the technology provide, for the first time, a closed system for handling information, and this development is considered to have fundamental importance for the future. Electronics is not essential to information technology; it bears the same relation to information that electricity bears to power — electronics is a convenient, but not unique, method of representation. It is expected, however, that over the next twenty-five years, almost all the important developments in information technology will be based on electronic techniques.

13

## DETERMINANTS OF THE FUTURE

In the past, it was the pattern of technology that largely determined the way computing developed. It is considered that this is no longer true. The technology available within the next five years will be more than adequate to generate great changes in the economic and social order. The sequence and timing of these changes will be determined not by technological factors, but by social and economic factors, and to establish a view of the future it is these that need to be studied rather than technological developments.

## MICROELECTRONICS

The development of information technology is based on the tidal wave of microelectronics, which is causing the cost of processing and storing information to fall to a level where these can be applied to everyday uses of information, rather than just to the specialised high-value applications currently satisfied by computers. These reductions in cost have been made possible by the increased complexity of integrated circuits. The technology is now reaching a level where the majority of current and foreseeable high-volume applications could be satisfied by single component systems, so that the demand for further increases in complexity is expected to decline, leading to a stabilisation of the technology. The future pattern of microelectronics is expected to show an increasing divergence between very complex circuits offering high performance at high cost, and high-volume circuits where the emphasis will be on minimum cost.

## THE MICROCOMPUTER

The microcomputer — a complete computer on a chip with processor, memory and interface — is seen as the key technological development. At present, semiconductor technology is inadequate to make an effective microcomputer, but this constraint will disappear in the next two or three years. The microcomputer is important because it enables single-component information systems to be achieved. These will be easier to design and cheaper to manufacture than multiple component systems.

The microcomputer is also important because it provides a new level of abstraction in the design process. Instead of designing at the circuit level using electrical parameters, or at the gate level using logical parameters, the designer will work at the system level using information parameters. As yet, the theoretical basis for such design techniques is lacking: there is no ready-made calculus at this level, as there is at the logical level, but such a basis may be expected to emerge from theoretical developments in programming languages.

## SILICON PRODUCTS

At present, there are a large number of different microcomputer architectures on the market. Market forces are expected to generate *de facto* standardisation leading to one or two preferred architectures, as has already happened with computers and minicomputers. There is a risk that *de facto* standardisation will cause the perpetuation of current architectures, even though they are unsatisfactory. Although standardisation of architecture will occur, it is expected that there will be increasing diversity of silicon products. These products will be obtained by integrating various standard system components like processors, memory, interfaces and peripherals onto single circuit to provide customised configurations for specific types of application, just as customised computer configurations are built from standard system components today. Many of these products will consist primarily of memory, with a microprocessor to provide some specialised forms of access, for example as in a silicon diary.

## PROGRAMMING LANGUAGES

Programming languages should be seen as the theoretical basis for information technology. There is room for considerable development in the structure and organisation of languages, particularly to provide open programming languages capable of intercommunicating with completely independent computer systems. Such developments will improve the design of computers and the efficiency of systems, but are unlikely to have substantial impact on programmer efficiency. Programming is intrinsically

15

difficult. Eliminating the mechanical aspects of programming will leave the programmer with the hard part, which is the formal expression and solution of a problem, so programming will become harder rather than easier. What technological development will do is reduce the need for the programmer to have any special expertise in computing.

## SOFTWARE

Technological developments are making the concept of the stored-program computer uneconomic. The high cost of software can best be reduced by selling multiple copies. This will lead to the development of a market for programmed information products, providing the functions currently performed by the computer. The implication is that the concept of the customer/programmer will disappear, while the software industry will move from a service to a product orientation, with a sharp change in competitive pattern. Since software products will have a high cost of development and a low cost of delivery, the competitive pattern will be similar to that of the semiconductor industry, with strong price competition, the need for large active markets, *de facto* standardisation and pressure to innovate.

## THE COMPUTING INDUSTRY

The structure of the computing industry has already changed substantially, with a large decline in the importance of the main-frame-computer manufacturers. This change is obscured by the visible nature of the mainframe computer itself. The developments in the technology are expected to accelerate these changes, leading to a destabilisation of the present computer industry under the competitive threat of plug-compatible products, mini-computers and programmed products. The existing capital investment in large-scale computers for data processing means that this area will be relatively slow to adapt and take advantage of the new technology, and will be overtaken in innovation and size by other markets.

## TYPES OF USE

Microelectronic technology facilitates the handling of information. This can come in a number of distinct forms. There is state information characteristic of control systems, textual information characteristic of human information systems, and spatial information characteristic of picture systems. It is to be expected that there will be increasing specialisation in the components used for handling the different types of information.

## CONTROL SYSTEMS

In the short term, control systems may represent the most important use of microelectronics, as existing mechanical and electromechanical control systems are replaced in consumer goods and industrial products. This will not, however, represent a large-scale market in the long term unless new ways are found to exploit the potential for control.

## INFORMATION SYSTEMS

The large-scale markets for the medium to long term are seen to be in the area of textual information. Most information currently used by humans is in this form, and the provision of a coherent information system for the processing, storage and transmission of textual information will have a profound effect on many aspects of society. The first phase of this is the introduction of word processing into the office. Because this can be done on a piecemeal basis and is readily cost justified, the introduction will be rapid. The availability of word-processing equipment will provide the infrastructure on which a unified electronic information system can be built, with electronic textual intercommunication within the office, and electronic mail between offices. Given this infrastructure, total electronic information systems will be highly competitive with current paper-based systems, and since they provide many more facilities there could be a rapid switch from the paper-based to the electronic-information society.

## THE ELECTRONIC INFORMATION SOCIETY

Microelectronics will make possible the electronic typewriter — a silicon replacement for the word-processing terminal and the visual-display unit. This device, with a page-sized solid state display and a keyboard, will have inbuilt text-editing and processing capability, cassette semiconductor storage and a serial interface to the information network. Its widespread use in the office will drive the price down to a level where it becomes a consumer purchase, just as already happened with the pocket calculator. As a result, it will be possible to extend the use of electronic information systems into the home, replacing the use of paper and providing capabilities for electronic mail (electronic newspapers), direct access to libraries and other forms of information, and silicon books and records.

## TELECOMMUNICATIONS

The major obstacle to the development of the information society will be the provision of adequate telecommunications capability. The present development plans of the British Post Office (BPO) and of postal, telegraphic and telecommunications authorities (PTTs) elsewhere are unlikely to provide suitable telecommunications capability until well into the twenty-first century. The future pattern of requirements for data transmission is expected to be quite different from that currently seen or forecast, the bulk requirement being for low-bandwidth transmission to a very large number of stations. Accordingly, the proposed pattern of data-communication development based on package switching may not be the most appropriate. The limitation will not lie in bandwidth or total capacity of the trunk system; the problem will lie in providing a suitable local digital-communication network enabling terminals to use a variable bandwidth capacity according to their needs.

## EMPLOYMENT

The most immediate consequence of the technology will be its impact on employment. So far, little capital has been invested in improving efficiency of workers in the information sector —

secretaries, typists, clerks and managers. The use of electronic information technology must be expected to have a marked effect on the level of productivity in this sector, and therefore on patterns of employment. The information occupations are thought to amount to 65% of the working population, so that even moderate improvements in productivity could bring about unemployment levels in the 10 to 20% region unless offset by compensatory increases in demand in these or other activities. By comparison, the impact of the technology on employment in the direct manufacturing activities, is expected to be relatively small. The overall consequences must be seen as comparable with the industrial revolution of 200 years ago.

## OTHER ECONOMIC CONSEQUENCES

In the medium to long term, there will be widespread use of microcomputers and other microelectronic products in consumer and capital goods. This must be expected to lead to a general shake-up of industry, with a change in the relative competitive position of many companies. In the same period, the impact of the technology will generate increased inequality between individuals and between companies. In the very long term, as these changes work their way through the economy, the overall level of productivity will increase and society will obtain the benefits.

## IMPLICATIONS FOR WESTERN ECONOMIES

There is a considerable risk that some countries will not exploit the technology as actively as their major competitors. Where a country fails to introduce microelectronics into its products, and fails to take advantage of the improvements in productivity that the technology will allow, its products may become non-competitive in the world market. This would have serious consequences particularly for major exporters such as the U.K.

In the case of the United Kingdom, from an economic point of view, it is far more important that the U.K. *use* the new technology than that the U.K. should *provide* it. The supply industry will amount to only 1% or 2% of the economy, whereas every aspect of the economy will be affected by the use of the technology.

In the medium term, the microprocessor will find its application in capital goods where it represents only a small part of the cost. As such, a U.K. semiconductor capability is not of great economic importance. In the longer term, as silicon products arise, access to a U.K. capability would become more valuable, and such access may also be required for strategic reasons.

Computer policy in the U.K. has concentrated on the main-frame-computer industry. This is a declining sector of the industry, and greater attention should be given to other sectors.

In the medium term, the software industry has considerable potential; however, as the market switches to software products, the U.K. will be at a disadvantage because its home market is not sufficiently large or active to provide a base for com-petition.

From a national point of view, the greatest leverage might be obtained by an active programme of investment in an advanced telecommunications system. This would provide the infrastructure for an information society and would generate a positive market for the information technology industry in the U.K.

**IMPLICATIONS FOR RESEARCH AND DEVELOPMENT**
Currently, much government R & D expenditure is directed towards the support of the computing industry and towards further developments of the technology. This should be redirected towards the use of existing technology in sectors outside the computing industry itself. The emphasis of such support should not be towards control applications but towards information-handling applications in the office and elsewhere. The role and responsibilities of the National Computing Centre (NCC) in this area should be expanded.

The government should consider reorganising the Government Research Establishments' (GRE) resources it has available into a single Institute for Information Technology. This would provide a focus for the subject and could play a valuable role by providing technical support services to governments and users in the following areas:
— technological forecasting and policy determination

- centre of expertise in semiconductor technology
- centre of expertise in languages
- monitoring of R & D in other countries
- R & D on standardisation
- use of technology in government.

# CHAPTER TWO

# STUDY APPROACH

## GENERAL COMMENTS
This study was sponsored by the Computers, Systems and Electronics Requirements Board (CSERB) of the Department of Industry. It was carried out by Iann Barron in conjunction with Ray Curnow and other members of the Science Policy Research Unit at Sussex University. The study was made during 1977.

## SCOPE OF THE STUDY
The study is specifically directed towards the technological implication of information systems and their consequences for government policy. Wider aspects of policy, and the social and economic implications, have been considered only insofar as they provide a necessary context for the discussion.

The study is concerned mainly with the next ten years. Major trends are expected to develop in this period, but not to become significant until subsequently. For this reason it has been necessary to examine a longer timescale. This is important, since it is considered that many of the changes that will occur after 1985 need to be taken into account when formulating policy for the next ten years.

The essential prerequisite to the formulation of a general policy must be the study and the thorough evaluation of policy alternatives. Such studies are particularly necessary in the case of computing because the situation is changing so rapidly that the formulation of a forward-looking policy is extremely difficult.

It is the opinion of the authors that it is social and economic factors, rather than technological factors, which will mainly determine the future development of information technology. This has two important consequences:

— Any attempt to predict future developments exclusively on the basis of technological considerations is doomed to failure.
— The formulation of proper social and economic policies in relation to computing is far more important than the formulation of a technical policy, and will have a much greater effect in determining whether or not the U.K. benefits from the potential of information technology.

## METHODOLOGY

We have sought to present the current status of the industry and the use of computing in as broad a manner as possible. This was done through discussion with knowledgeable individuals and by the evaluation of published information. Projections into the future have been based on this information and have been derived by consideration of analogous situations, case studies and analytical discussion. Particular emphasis was placed on the consideration of relevant historical examples of technology diffusion. The overall technique is best described as one of intelligent extrapolation, which obviously is both subjective and imaginative.

It is apparent that the pattern of development will be determined mainly by non-technological factors; these have been considered, but not to the depth thought desirable.

This methodology has led to conclusions rather different to some of those which are generally held. The authors acknowledge the help and assistance that they have received from many individuals but accept full responsibility for the forecasts that have been made from the information collected.

## STRUCTURE OF THE STUDY

The results of the study are presented as a sequence of chapters considering different aspects of the future. It must be understood that this organisation is purely a matter of presentation and that there is a high degree of interaction between the various topics discussed. Although, for convenience, as much as possible of the presentation is serial, in reality there is no such logical development, and many of the conclusions are based on highly involved analysis and argument.

Within the report the following terms have been used to indicate approximate time scales:

| | |
|---|---|
| Short term | up to 1980 |
| medium term | 1980-85 |
| long term | 1985-2000 |
| very long term | 2000 and beyond |

To provide an overview, Chapter 3 gives a general discussion of the impact of the technology and the changes it will bring about. This chapter is based on the more detailed discussions which follow in the rest of the report.

Within the report, some sections are italicised. These sections represent conclusions, recommendations or summaries justified by the preceding text.

# CHAPTER THREE

# TOWARDS THE INFORMATION SOCIETY

This chapter gives a general overview of the developing role of information technology in the economy and of the major consequences that are foreseen.

As yet, the unit price of computing equipment has been high, and this has restricted its application to large-scale organisations. The consequence is that even though the computer now plays a central role in the operation of the economy, the general public has little awareness of its importance, or of its potential to influence the future.

The thesis of this study is that the developments of the technology, which are discussed in Chapter 4, have the capability to reduce the unit cost of computing devices to the extent where their use can be disseminated widely through industry and the home. As a result, the already crucial importance of computing to the economy will be greatly magnified. The further consequence is that the computer may in the future have a far more direct impact on the general public, and conversely that the general public may exercise much greater influence over the development of the computer and its pattern of exploitation.

# THE INFORMATION ECONOMY

Since its development thirty years ago, the computer has come to occupy a central role in the economy. Nevertheless, its range of applications has been severely limited by its high cost. With the advances in the technology that are now occurring, the range of applications and the importance of the technology to the economy and to the individual will greatly increase.

## The importance of computing

In its relatively short life the computer has grown to be a major factor in our society. Military capability is now almost totally based on the use of computer-controlled systems, which have created a new degree of responsiveness and integration in military organisations. In the same way, major business operations have become completely dependent on the use of computers to plan and manage their activities.

The extent of the dependence on computers by modern society is demonstrated by a recent American study which considered the consequences of a total withdrawal of computer facilities, for example by strike action. In the order of events, the more important consequences are:

| | |
|---|---|
| — loss of military security: | due to failure of surveillance systems and the inability to launch counter offensives. |
| — disruption of communications: | much of the telecommunications network is already vulnerable. Air transportation would be impacted almost totally, and there would be some impact on rail transport, particularly the distribution of freight. |
| — loss of internal security: | due to disruption of communication and police control systems. |
| — disruption of industry: | Electric power generation and distribution is largely dependent on computers, and could not be maintained at a viable level. Within industry many operations, |

|                      | particularly in continuous processes, |
|----------------------|----------------------------------------|
|                      | are critically dependent on computers. |
| — social disruption: | many companies would be unable         |
|                      | to calculate salaries and the banks    |
|                      | would be unable to transact the        |
|                      | payments.                              |

Overall, the disruption caused by the loss of computer facilities would be even greater than the disruption caused by the loss of the electricity power-supply system, which is usually regarded as the most central element of our world system. For the present, at least, the loss of computer capability is an event of relatively low probability, because computer capability is widely distributed throughout the economy and its control is not centralised.

## Information technology

The high cost of computing has limited its application to a few situations where information has high value. As the cost of computing is reduced, so the range of use can be extended; the ways in which this might happen are discussed in Chapter 6. So far, computers have been concerned almost totally with numerical information. There is no reason, however, why the computer should not be used to handle other forms of information.

Most of the information we use is not numerical, but textual, and it is the extension of computing techniques to textual information that is expected to bring the greatest change in the future.

In the past, information has been handled by a variety of techniques. The most commonly used storage medium is paper; the most common transmission systems are post and telephone; input-output equipment ranges from quill pens to the typewriter; and the most common processing device is the human brain.

The developments in microelectronics provide, for the first time, a closed system for handling information in terms of the single medium of electronic signals. Using this medium, it is going to be possible to handle information at a price which is highly competitive with present-day alternatives. It is already cheaper to transmit and store information electronically than to use paper and the telephone. The only restriction to exploiting these advantages has been the cost of capturing information in electronic form. The latest developments in microelectronics

27

mean that within the next ten years it will be cheaper to capture information electronically using an electronic typewriter than it is to capture information using a mechanical typewriter. At that time, completely electronic systems not only become possible, they become directly cost competitive with the present-day alternatives while providing many advantages due to the flexibility with which electronic information can be altered or retrieved.

These developments are already being recognised in the discussion of the convergence between communications and computing. It is becoming less useful to discuss the evolving technology in terms of these traditional divisions. Neither the present reality of communications nor that of computing is likely to play a major part in the future of the technologies. Instead it is the underlying concepts like the switching of information and the use of programmability that will be the significant features. It is for this reason that the subtitle of this study is *Forecasting the Effects of Information Technology* rather than *The Future of Computing*, or some such title. It is to emphasise that the current purposes and implementation of computing may not be fundamental but may only be aspects of some more important underlying developments.

The term 'information technology' has been chosen to describe this subject, because it emphasises that the basic area of concern is information, and that for the first time we have a coherent technology to handle information. So far as possible this term has been used throughout the study, although on occasions the term 'computing' has still been used in order to assist comprehension.

Information technology includes the processing of information (which is currently performed by computers or manual methods), the storage of information (which is currently largely non-electronic) and the communication of information (which is currently performed by voice, telecommunications and postal services). Information technology is emerging as a systematic discipline because the development of microelectronics provides a coherent basis for all its aspects. Electronics itself is not fundamental to information technology; it bears the same relation to information that electricity bears to energy — it is a convenient, but not unique, method of representation. It is expected, however, that in the long term, it is electronic information which will be the dominant factor, and that the important developments of the next twenty-five years will be due to the exploitation of the electronic representation of information.

The evolution of information technology as an identifiable discipline will have direct consequences for the organisation of the subjects concerned. At present, the various aspects of information technology are regarded as separate subjects, like electronic engineering, communications and computing science. Associated with these subjects are a plethora of structures — university departments, research organisations and professional institutions. As the convergence of these subjects continues, so the degree of overlap and interaction between the subjects will increase, causing increased confusion until it is resolved by the revision of the associated structures. The early reorganisation of these subject areas would have considerable advantage, both by reducing duplication and by bringing together the boundaries where active developments are occurring.

### The theory of information

The concept of information is ill understood. One dictionary definition of 'information' is 'intelligence given', while the definition of 'intelligence' is 'information communicated'. As yet, there are no adequate theories to describe information. Shannon's theory, while relevant to the communication of coded signals, provides little insight into the nature and relevance of information. Like early theories of heat, Shannon's work captures one aspect of the subject, but clearly excludes much of importance. The state-change theories of Petrie and Holt likewise only capture a limited aspect of the information content. The need for a better theoretical basis for information is urgent, and it is to be expected that advances in this area will be promoted by the emergence of information technology as an important subject.

Information has to do with the representation or mapping of one system by another. The system represented is the universe of discourse, the representing system is the language, and the mapping is the semantics. Within such an interpretation, it would seem that the analogy to information is not the entropy of a system as developed by Brioullin and Shannon, but energy

29

itself. Indeed, the analogy between information and energy is very close at many different levels, and needs to be studied.

**Information as a resource**

Like energy, information is a basic resource which is essential to the operation of a country's economy. The provision of a cost-effective information system is just as important to the future of the U.K., for example, as is the provision of a cost-effective energy system.

The importance of energy to the British economy is recognised and is a matter of active political and social concern. In the case of energy, there is:

— a Secretary of State with Cabinet status, responsible for planning energy policy
— a Department of Energy to manage energy resources, and develop and implement policy
— a variety of active large-scale research programmes into short- medium- and long-term aspects of the provision of energy, including programmes like the fusion project which are directed towards the twenty-first century
— public awareness of the importance of energy, and public debate about policy issues like the reprocessing of nuclear fuel at Windscale.

Contrast this with the importance so far attached to information technology. And yet over the next twenty-five years:

— The scale of public investment in information technology through the provision of telecommunications capability will be comparable with the scale of investment in energy systems.
— The scale of corporate investment in information technology through computing and developments like word processing will be comparable to the scale of investment in energy systems.
— The scale of home investment may also be comparable if adequate information-communication services become available.
— The harm to the individual caused by the misapplication of information technology could well outweigh the potential risk from nuclear hazards.

Unlike energy technology, information technology is a new and rapidly developing area, and the societies which succeed in managing and exploiting this new opportunity will have a major advantage over their competitors. It must be a matter of the greatest concern that there is so little public awareness of, or political emphasis given to, this key subject.

## FUTURE SCENARIO

A considerable amount has already been written about the future information-based society. Much of this is fanciful and fails to bring together the major constraints of technological limitations, investment requirements and slow social adaptability. The following scenario has been based on the information acquired during the course of our study, and it attempts to take into account these major constraints: in particular it is assumed that whenever possible the technology will be used to replace existing products, rather than to create new patterns of demand. Such new patterns will occur, but they are far less predictable, and in any case are unlikely to be significant before the very long term.

### Communications systems

The basis for an information society must be a coherent and reliable communications system. Such a system will look far more like the electricity-supply system than the present telephone network. The house and the office will be wired for information, with information sockets like three-pin power sockets for electricity. The system will use the ring-main principle, with information transmitted serially on a cable, the information socket being an active device allowing information to be inserted or removed from the ring main. Information from the home system will flow out to the external network through a junction box, which will provide physical isolation for the external system and metering, the charge for using the external network being based directly on the number of bits transmitted and the number of bits received.

Within the home or office, there will be a variety of different products which could be plugged into the information supply. Unlike the present-day telephone, which is rented from the

31

telephone company, these devices would be bought by the house-holder to meet his personal requirements, just as, at present, he can buy a variety of devices from hi-fi's to waste disposers, which can interconnect with the electricity supply. The opening up of the market to independent suppliers would not only have the effect of stimulating the market for information products, but would also release capital investment for the information authority to concentrate on investment in the information network itself if required.

The communications network will require digital working right out to the subscriber stations, and the local-level distribution systems would again be based on the use of a multiplexed ring-main system to enable variable information bandwidths to be drawn off at each station as required. The current proposed pulse code modulation (PCM) voice network could form the basis for the trunk distribution system, but a completely new investment will be required to provide the local communications capability. The late availability of an adequate data-communications network is likely to be the major investment obstacle to the information society.

Investment in such a network should be seen as having economic importance comparable to that of the investment in the road network or in the electrical supply system. Provision of an adequate data-communications network is perhaps the most effective way in which a country will be able to obtain an advantage over its competitors: it will both create the environment for the application of information technology and also provide a direct stimulus to the information-technology industry.

### The office

The largest changes are expected to take place in the office. Most office work is directly concerned with information and is amenable to improvement using information technology. At present, the office uses paper-based information systems. It is expected that these will be completely replaced by electronic information systems, eliminating the mechanical typewriter, the filing system, the photocopying machine and postal services. This change is expected to occur over a relatively short period of ten to fifteen years, because the electronic systems will be directly price-competitive and will provide enhanced facilities. The parallel operation of both paper and electronic systems is very much more expensive, and it cancels out many of the potential benefits: it is not difficult, for example, to imagine a secure, read-only, electronic image as replacing the 'original document' required for legal security.

Within the office, the key device will be the electronic typewriter. This is the direct electronic analogue of the conventional typewriter, with a page-sized information display and a keyboard input. It might very well have similar dimensions to a pad of paper, and will be connected direct to the office information ring main. The electronic typewriter will enable information to be entered as on a conventional typewriter. It will also have built-in word-processing capability, enabling the formulating and editing of text to take place, and some limited calculating abilities. Finally, the electronic typewriter will have the capability to transmit and receive information through the information ring main.

The information ring main would not only interconnect the electronic typewriter, it would also interconnect the office computers performing specialised functions like sales-ledger accounting, market forecasting, and stock control. In addition, it would support one system dedicated to the filing of information and another dedicated to controlling message interchange. Finally, the information ring main would connect to an information exchange, controlling communication with the external public-utility information system.

The basic electronic typewriter would be used for all purposes in the company, and would be more widely disseminated than the present typewriter. As today, secretaries would be the primary method for high-volume input into the system, but managers and others would also have their own terminals, enabling them to retrieve information from their files, to receive letters and memos and to provide a method for interacting with computers within the company or elsewhere.

Intercommunication between companies would be direct, using electronic mail, probably with overnight delivery to balance the load on the communications system. This would apply not only to letters and similar information, but to ordering and invoicing, which could be performed directly on a computer-to-computer basis, given the necessary agreement on information formats and significance.

Within such an environment, it has been suggested that the office would be eliminated, since it will be possible to access all the relevant information from a home terminal. This is not considered a likely development in the medium or long term, because it ignores the fact that much of the interaction within a company is not carried out on a written basis, but depends on personal contact, and also that the act of going to work is in part a social function.

The introduction of the electronic office would make the operation of offices considerably more efficient, reducing the overall need for office staff; this would include managers as much as clerks. The use of word processing should lead to a direct reduction in the demand for typists, but it is not envisaged that the role of the secretary will be eliminated — although it might change considerably as the routine tasks like filing are automated and managers also have access to the facilities through their own terminals.

### The factory

The impact of information technology on the factory is expected to be far less dramatic than the impact on the office. As is not the case with the office, there is already a high degree of investment in the factory, so the room for improvement in productivity is reduced. Also, the primary role of information technology is to handle information, and while information may be relevant to the manufacturer of goods, it is not so important as it is in the office. This is not to say that the qualitative restructuring of factory organisation may not outweigh any quantitative improvements.

Over the past fifty years there has been a steady increase in the number of people involved in the information sector of the economy, as opposed to agriculture or manufacture. Many forecasters have suggested that this trend will be accelerated by the availability of electronic information systems. The increase in the information sector can be accounted for largely by the failure to improve the productivity of this sector in comparison with agriculture and industry, the increase in demand for information products being a secondary factor. An improvement in productivity in the information sector might be expected to stimulate a a further increase in demand, but this should be more than offset by the improvement in productivity, so that in total the information sector should decline in relation to agriculture and industry. Thus, the paradoxical effect of the information revolution could be to increase the proportion of the population engaged in industry, agriculture or services other than information services.

The automation of existing control operations in factories could proceed quite rapidly. The low cost of microelectronics will allow direct replacement of mechanical and electromechanical control systems, whereas in the past the high cost of the computer has necessitated radical changes to the design of plant, before its introduction could be economic. Microcomputer control systems can be used to control individual instruments or machines, and it is expected that a pattern of piecemeal automation will arise where intelligence is added to existing production processes in an ad hoc manner. From the point of view of the worker, such an approach could be attractive, since it is likely to provide him with tools and instruments which enhance his skills and make the job more interesting, in contrast to earlier phases of automation which have reduced the skill level of the individual. The human is unlikely to be eliminated by the greater introduction of micro-electronics, because it is not expected that it will be possible, even in the longer term, to achieve systems with the degree of adaptability provided by the human. The perceptive and mechanical skills of the human are based on an extremely complex processed representation of reality, and it is not likely that micro-electronic systems will be able to achieve this capability in the foreseeable future. Some further elimination of human involvement in the manufacturing process is possible, but this would only be achieved in the same way as in the past — by eliminating the areas of adaptability and tolerance, which demand human skills. As such, the information revolution does not impinge directly on the problem.

The risk with increasing automation is not that the human content in manufacturing is reduced but that the skilled content is eliminated, leaving one or two types of job where physical capability alone is essential. The example of some chemical processing plants — where the whole of the manufacturing is controlled by a handful of technologists, but where the final

35

output stage of loading the chemicals into bags and on to trucks or wagons must still be carried out by manual labour — shows the dilemma of the technology: that it may further divide working jobs into two classes, one requiring intermittent and intelligent action, the other requiring continuous physical labour.

The replacement of existing control systems in factories is unlikely to have a significant effect, because these control systems, based on mechanical or electromechanical parts, are very simple. Indeed, most microelectronic control systems will have a great deal of excess capacity as they directly replace the existing control systems. The full exploitation of microelectronic control must be expected to take a long time to develop, because it involves acts of creative intelligence to identify new ways in which to exploit control capability.

There is another reason why the impact of microelectronics on the factory is expected to be less dramatic than often forecast. The productivity of the manufacturing worker has already been greatly increased by a variety of techniques, and he now represents a considerable capital investment. The investment of further capital, in the form of electronic control systems, can, therefore, only be expected to have a marginal impact on his performance. Finally, the number of people directly involved in the manufacturing process (that is, excluding foremen, supervisors, administrative staff and management) is relatively small — less than 20% of the working population — so the effects of microelectronics in the manufacturing sector are unlikely to be very great in overall terms.

**The home**
It is not expected that the television set will become the information centre of the home, except in the short to medium term. Instead, it is expected that information will be distributed and be accessible throughout the home. The development of low-cost display systems will mean that it will be economic to provide multiple terminals at which information can be accessed or inserted, and they will be used in socially more convenient positions, like the kitchen and the workshop, rather than obtruding in the leisure area.

Neither is it expected that every home owner will become an

expert programmer, slave to his home computer. The concept of a single computer controlling the various aspects of the home, which the owner programs to meet his own requirements, seems implausible. It is based on an extrapolation of the present constraints of the computer in terms of cost and mode of use. Programming is a difficult task, which will be made harder, rather than easier, as the mechanical aspects of programming are eliminated by improved software-development techniques. The important products for the future are not programmable products, but programmed products, and the householder may be expected to be a considerable consumer of these without being able to tell a 'for statement' from a 'do loop', just as he can already operate a pocket calculator with no knowledge of the way the numerical algorithms are programmed inside.

In the short to medium term, many home devices, from television to cookers, may be expected to incorporate microcomputer controls. Such products will be developed independently by separate manufacturers, and will be conceived as free-standing services unable to intercommunicate, except with a human controller. In the longer term, when a home information-distribution system emerges, it may be expected that the various household devices will be developed to access and transmit information through the distribution system, so that, for example, the cooker can be turned on remotely by telephone, the gas meter can be read remotely, and the television can be piped around the home.

The variety of programmed products in the home may be expected to be large, although the average householder may be unaware of their existence. At the moment, the average household has at least some fifteen fractional horsepower motors, hidden in such products as the food mixer, the washing machine, the electric shaver and the model railway. Indeed, the motors are so well hidden that the average householder can only remember some four examples if he is asked. The role of the microcomputer in the home will be very similar; it will be embedded in products which have a specific function, and which are used without any need for specialised expertise.

In the short to medium term, the number of microcomputers in the home may well be comparable to the number of electric motors. Most places where the electric motor is used have an associated control system which could benefit from the use of a microcomputer. In the longer term, a variety of new uses may be expected to arise, which will cause the number of microcomputers in the home to rise considerably. The provision of an information distribution system around the home will need intelligent wall sockets, each containing a microcomputer. To access information on such a system, electronic terminals and electronic notepads will be widely used. Within the home, information will be stored on semiconductor devices to replace paper, voice cassettes and video tape. It is to be expected that a whole range of consumer information products will arrive which will use silicon in one form or another.

While the home computer itself may not be a sensible concept, it is likely that a number of the facilities currently available on corporate computers will become available to the householder in the form of specialised information products, such as the home accounting system, scheduling and planning systems and information-storage and retrieval systems. Such devices will be sold as identifiable products, to be purchased by the householder and plugged into his information-distribution system.

The development of information technology is expected to have its most radical effect on the way the householder obtains information. Some of the possible developments are:

— selective television and radio, the viewer ordering his requirements through the telephone system, rather as he can currently select a library book
— direct access to libraries
— printed books and plastic records replaced by semiconductor information cassettes
— newspapers transmitted by telephone or radio
— selective transmission of information encouraging far more specialised information services (for example on sports) to become available
— selective control of advertising by the consumer, perhaps leading to rather different patterns of selling
— remote shopping and ordering
— home electronic mail

Not only will the technology enable there to be much greater freedom in the transmission of information, it will also make much greater selectivity possible and it is this aspect which is expected to generate the most change in the way that the consumer uses information.

Obviously such change would have major cultural impact, not least in use of and attitudes to leisure. Certainly, choice of access to provided information, whether for education, entertainment or for economic interaction, is enhanced. One possibility is that distinction between work, leisure and other activities may become blurred.

## THE INFORMATION REVOLUTION
The development of information technology is manifestly analogous to the development of mechanical power at the time of the industrial revolution. Will there be a new information revolution creating as fundamental an effect on society as the introduction of the steam engine?

### The industrial revolution
There are considerable parallels between the present situation and the industrial revolution two centuries ago. Prior to the effective development of steam power, there was a period of well over 100 years in which the need for mechanical power became more apparent, and a variety of intermediate technologies were used. The original development of steam power at the end of the seventeenth century was in the form of the single-acting beam engine, which was used almost exclusively to pump water. As such, it had considerable economic significance, in that it enabled coal to be mined at much lower depths and made the canal system practicable. Improvements to the design of the steam engine in the third quarter of the century, and the crucial invention of the crank and the planetary gear, which enabled the steam engine to provide rotary power, allowed its use in a much wider range of applications. One of the earliest of these applications was in iron making, to improve the mechanical strength of the steam engine itself; another was the commercial use of the steam engine in brewing. Indeed, the first steam engine to be installed in London was at a brewery. Thereafter, the applications of the steam engine occurred in waves over a period of seventy years as major

sectors of the economy were affected — manufacturing industry, rail transport, agriculture and, finally, shipping.

The analogies with the growth of information technology are obvious. In the nineteenth century there was an increased demand for information aids, which culminated in the development of the tabulating machine. The development of the computer in the third quarter of the present century is a direct consequence of this demand, and has led to the development of pervasive microelectronics which has potential applications throughout all sectors of the economy. It is not unreasonable to expect that the analogy will continue, with exploitation in different sectors of the economy occurring in waves over a long period. To be speculative, these waves might be: textual information systems, picture processing, mechanical automation and, finally, artificial intelligence — although only the first of these comes within the purview of this study.

The analogy can be pursued still further. The steam engine of 200 years ago was a very large device which was housed in a specially designed area, and constrained the organisation of the factory. Power was transmitted from the central steam engine by huge spindles made from tree trunks, and was picked off by belts which drove a variety of machines. Such a contraption sounds uncommonly like a conventional computer driving many user programs through its operating system.

The significance of the steam engine was that it enabled a step-function increase in productivity. The consequences of this were enormous. For the fortunate, it brought wealth and power; but for many it caused social disruption, greater poverty and a life sometimes even more abhorrent than the rural rigours they had previously suffered. The consequences were not due to the steam engine itself, but to the inability of the economic system to absorb the changes, and to side effects like the movement from rural to urban areas.

### Kondratiev long waves

The cyclic behaviour of the economy is well known. However, apart from the short economic cycles, which frustrate management and government, economists have identified much longer cycles lasting fifty to sixty years, which have been called Kondratiev long waves. So far, three such major waves have been described. According to Schumpeter's interpretation, the first of these was associated with the introduction of the steam engine and lasted approximately from 1780 to 1840. The second was associated with the introduction of railroads and lasted approximately from 1840 to 1890. The third was associated with the introduction of electric power and the automobile and lasted from 1890 to 1940.

The characteristic of the Kondratiev wave is that in the first phase of the wave, the economy speeds up. In the second phase, there is a reversal of this trend, leading to a major depression in the downswing. Associated, therefore, with the Kondratiev long waves are the great depressions of the 1930s, the 1870s and the 1820s.

Economists differ in their interpretation of the Kondratiev long waves, but a simplistic description may be obtained by considering the effect on the economy of a major potential increase in productivity. In the first phase, the new technology is being generated. It has little impact on industry as a whole, but creates a small, rapidly expanding sector which is highly profitable and attracts heavy investment. In the second phase, the effects of this technology are disseminated through the economy.

Thus, while there are undoubted long-term benefits from major improvements in productivity, the consequences on a shorter time scale have often been to the disadvantage of a large part of the population. These consequences can be directly traced to the imperfect operation of the economic system under dynamic change, and are in no way related to the technology itself. Two other effects of the operation of the economic system should also be noticed:

— The changes lead to considerable inequality in the distribution of benefits from the productivity improvement. Some groups, particularly those associated with the exploitation of the technology, benefit greatly, while others suffer disproportionately, either because their jobs are eliminated by the changes that occur, or because they are in no way affected by the technological change, so that they cannot participate directly in the long-term benefits. Again, these inequalities are not permanent; they are eliminated as the economy stabilises.

— The same effects apply to companies and to whole industries, since the raising of the level of demand is not
uniform for all products. Even if the level of productivity
were increased by a factor of two, the demand for wrist-
watches, for example, might not increase by the same
factor; if prices fell sharply, some part of the increased
spending power might be channelled into other avenues.
Some companies will benefit disproportionately from the
change in productivity; other companies may find the
demand for their products reduced, or even eliminated
by new patterns of usage.

**The information revolution**
Currently, world economic growth is slow and unemployment
levels are rising: this pattern looks suspiciously like the fourth
Kondratiev wave. The technological change associated with
this wave is electronics generally, and information technology in
particular. The 1950s and 1960s saw a boom in industry with the
rapid growth of consumer electronics and computing. Now, with
the pervasive microprocessor, this technology is affecting much
of the rest of the economy. Since there have been no basic changes
to the economic order, there is no reason to expect that the response of the economic system will differ significantly from the
previous Kondratiev wave. There are two provisos to this, however,
although it is difficult to assess their importance:
— Since Keynes we have begun to understand the feedback
effects of unemployment, and this may help us manage a
declining economy more effectively.
— For the first time, political power is far more widely
disseminated through the economy, rather than being the
prerogative of a minority who also control the industrial
wealth. The consequence is that it will be possible for
those sectors of the economy that are harmed by the
changes to exert far more leverage, and perhaps to control
or even stop the rate of change. Since the interests of
most individuals are perceived in the short term, this could
delay or prevent the undoubted long-term benefits that
would accrue from exploiting the technology.

# CHAPTER FOUR

# TECHNOLOGY AND FORECASTING

This chapter reviews the various aspects of information technology. The forecasts are based on the availability of new technology, the primary generator of technical change being the development in microelectronics. Developments in other areas like communications and software are seen mainly as a response to the pressures from microelectronics, rather than as being due to separate generators within their own areas.

The forecasts take into account the interaction between the various areas, and as far as possible also take into account the commercial incentives to pursue some lines of development rather than others which may have equal technological merit. Thus the pattern of development envisaged in this chapter presupposes the developments in industry structure and the usage of information technology discussed in Chapters 5 and 6.

The forecasts as presented are totally surprise free, and represent the working through of known technological concepts and capabilities. It is considered that any surprise innovation would take ten to fifteen years before becoming a significant factor in the pattern of development of information technology.

## THE VALUE OF FORECASTING

There is a considerable time-lag between the development of technology and its widespread application. Major ideas like the transistor or the high-level language can take ten years to go from concept to commercial exploitation and a further five years before they achieve widespread user impact. Even minor innovations like the matrix printer or virtual storage can take five years to develop and a further five to achieve widespread user impact. Thus, it should be possible to predict the general trends in computing over at least the short and medium term without recourse to any form of technological forecasting.

In practice, forecasts of the way computing will develop have been relatively unsuccessful. This would seem to be not because the basic technological forecasts have been wrong, but because the implications of the technology have not been identified, and a variety of non-technological constraints have not been considered.

It is possible to classify forecasting techniques into four types:

— Trend analysis and extrapolation. This is the basis of most market forecasting and therefore of the planning on which the computing industry itself is based. As such it has some degree of self fulfilment. While often good for the short term, it is of much less value in the long and medium term. Its short-term validity may be improved if simple trend extrapolation is modified by information obtained from the decision makers. The failure for all but the short term indicates the short-term view of most commercial operations.

— Scenario approach based on the Delphi techniques. The Delphi technique involves the polling of experts and suffers from two deficiencies as usually practised. The original formulation of questions can be very ambiguous and may bias the results, while the experts often fail to be experts, or if knowledgeable are expert in current technology and not in its future potential.

— Scenario approach using science fiction. Science fiction may be characterised as individual forecasting with the relaxation of one or more known scientific constraints. This technique has achieved some good long-term results, and its success may in part arise from the fact that forecasts

44

are based upon perceived goals rather than technological opportunity. It must also be due to the wide variability of such forecasts, which nullifies their value except in a post hoc analysis.

— Goal seeking definitions of the future, based upon deliberate positive or negative actions. Such normative forecasts seem the most attractive for the medium term, since they take into account positive external interaction which will affect the course of the technology. They may also make use of extrapolation and scenario techniques.

We have examined each of these various approaches in turn, by considering a variety of forecasts made in the middle to late 1960s about future developments in computer technology.

The main forecasts examined were the Diebold study of computer markets to 1975, the *Quantum Sciences* forecast on time sharing, the US Navy on Data Processing Developments and the Canadian Information Processing Society on Computer Patterns in the 1970s for Canada. All the forecasts went badly astray. Although the importance of the integrated circuit in reducing the cost of computing was identified, the possibility that this would create a new market for minicomputers and change the pattern of usage was not explored. The primary mistake was to see the future development of computing in terms of the then modish techniques of multiprogramming and time sharing. This led to an overestimate of the growth of large-scale computing and its associated techniques, such as the demand for telecommunication facilities.

This suggests that much greater emphasis needs to be placed on the fundamental changes that the technology will introduce, and on the ways in which these will be modified by social and economic pressures. Such a programme is not easy. Even given substantial information on the operation and planning of a large company like IBM, it is difficult to predict their response to changes which threaten the basis of their current operations. Equally, the relative failure of government computing policy in France shows that the adoption of normative techniques is not of itself sufficient, if the underlying assumptions are wrong.

It would seem that, in the past, the development of computing has been limited to a considerable extent by the available technology. The creation of new applications and the opening of new markets has followed technological innovation, and prediction of technological innovation has therefore been a good basis for predicting developments in the use of computing. The view is taken that this situation no longer pertains. With the latest developments in microelectronics a vast reservoir of technology has been opened up and it will be a long while before this reservoir is fully exploited. Further advances in the technology are thought unlikely to have a substantial effect on the way information technology will develop in the medium to long term.

What, then, are the significant factors that will shape the way information technology develops? The most important, in our view, are the diffusion of knowledge about the technology and the social responses which the adoption of a pervasive technology will generate.

We have concentrated in this study on the technological developments, and have attempted to translate them in terms of the potential uses of the technology and the pressures these will generate on the structure of the computing industry itself. This is only a partial view; much work is necessary on the social consequences of this technology (considered to some extent in Chapter 7) to reflect back the social interactions on the developing pattern of information technology.

Is forecasting then useless? We argue not, because there are lessons to be learnt with strong implications for policy:

— By discussing the technological potential it is possible to delineate the boundaries within which developments will occur and decisions will be required.
— By identifying the scope of the implications of the technology, it is possible to delineate the area which will be affected by policy and where policy needs formulation.
— By indicating the alternatives it is possible to provide a basis for policy which may be robust against the uncertainties of the future, and also to provide a better basis for public debate.

The approach to technological forecasting adopted in this chapter and in the rest of the study is complex. While account has been taken of the various forecasts that have been made elsewhere, we have attempted to identify the basic technological determinants and then to consider their consequences in a structured way, both in terms of the changing pattern of use and in terms of the changing pattern of supply, so that more general conclusions can be drawn about their impact on the economy and on social problems. Our approach is thus a combination of all four of the techniques described above, and may perhaps most suitably be described as 'intelligent extrapolation'.

## MICROELECTRONICS

The current trend towards increased complexity has been based on a demand for increased functional capability. The technology is now reaching a level where this demand is expected to decline, while further advances in complexity will require much greater capital investment. As a consequence, the pattern of microelectronic development is expected to diverge, with a continuing slow cost reduction in high-volume devices, and the emergence of very complex devices offering high performance at high cost.

### Trend Pattern

The development of computing has been rushed along by a torrent of microelectronic technology. The simple rule that the capacity of an integrated circuit will quadruple every two years has provided a basis for many predictions about the computing industry. Indeed, as soon as forecasts have moved away from this simple rationale to consider more complex interactions, they have proved increasingly fallible.

The steady advance in microelectronic capability has been based on the increased complexity of the integrated-circuit chip. While the cost of an integrated circuit has at least remained constant, if not increased, the number of transistors on the chip has increased exponentially, leading to dramatic reductions in the cost per function. These developments have been of direct benefit to the user, since most systems have been of such complexity that they require many integrated circuits for their implementation. The increasing capability of microelectronics has, therefore, enabled the user to build systems with fewer integrated circuits, thus reducing the semiconductor cost, and even more important, the support costs of packaging the electronics.

The technical optimists, who include most of the semiconductor industry, present a picture of continuing advance in microelectronics. This optimism is epitomised by the performance of the industry in relation to semiconductor storage, where there has been a change of three orders of magnitude in capacity, and hence in cost per bit, over the last ten years.

In practice, the improvement, while outstanding, is not as good as this simple picture implies. A 16K storage device like the 2116 is *not* sixteen times as complex as a 1K device like the 1103, because there are overheads to access and drive the array. The improvement in definition, which determines the minimum size of the transistor and connecting track, has only been a factor of three in ten years, leading to a direct increase in circuit density of a factor of ten. Other improvements have been obtained by an increase in chip size, simplification of basic circuits like the storage cell, and more complex processing techniques like double poly-silicon which enables the size of the storage cell to be reduced still further.

**Technical developments**
Looking to the future, it is necessary to ask whether the expected increase in capability of the integrated circuit can continue, and whether adequate demand exists for increased circuit complexity.

The increase in circuit complexity is a function of three factors:
— improved definition leading to smaller components
— increased chip sizes
— novel circuit and process techniques
Each of these requires separate consideration.

Further improvements in definition are possible. Physical limits on semiconductor behaviour would allow a further reduction of two orders of magnitude in physical dimensions, but fabrication difficulties will impose earlier limits. Conventional photolithographic techniques are reaching their limit as the resolution nears the wavelength of visible light. It is possible to go beyond this limit by using X-rays ro replace visible light and creating the necessary masks by electron-beam machining. There is, however, the further difficulty of aligning successive exposures of the wafer

48

with sufficient accuracy. Until this problem can be overcome, resolution is likely to be limited to about 3 micron (state of the art is 6 micron). The solution to the alignment problem would seem to be the development of new devices in which the critical dimensions can be specified by a single lithographic step, together with auxiliary techniques like ion implantation.

It is possible to overcome the limitation of the photolithographic technique by the use of direct electron-beam machining, with all the processing done *in situ* to eliminate alignment problems. It is this technique which has led to the view that circuits containing a million devices will be available in the mid 1980s. Direct electron-beam machining is intrinsically expensive, since it is a one-dimensional process, as compared to the two-dimensional approach of photolithography. With electron-beam machining, the microstructure must be created by a raster scan, and to achieve a resolution of 1 micron on a 10cm wafer requires $10^{10}$ pixels. This represents an extremely long exposure time tying up expensive capital equipment.

The overall size of the chip is limited by two factors:
- As the chip size is increased, the yield falls, because of imperfections, so there is an economic limitation on chip size.
- As the chip size is increased the power requirement increases, leading to more complex packaging in order to maintain the chip temperature at an acceptable level.

Imperfections may be regarded as being distributed randomly across the wafer. This leads to a situation where the yield in terms of good circuits shows a sharp decline after a critical size is reached. As production technology has improved, this critical size has increased slowly to its present area of around 35sq.mm. Since there is no yield advantage in making chips with a smaller area, most new semiconductor devices cluster around the maximum economic yield size. It is to be expected that improvements in production technology will continue, leading to an economic chip size of about 80sq.mm. by 1980.

Dissipation is more of a problem. The chip temperature must be maintained at a reasonable level to prevent degradation. To a first order, the power dissipation is largely independent of the definition of the circuit, but is characteristic of the type of circuitry used and is proportional to the area. For present packaging techniques in a still-air environment, acceptable dissipation is limited to 1W or less, which implies a limit on chip area of 50sq.mm. for nMOS circuits or 500sq.mm. for IIL circuits.

The final areas of the improvement are the interrelated aspects of circuit design and process technology. The move from pMOS to nMOS and the introduction of IIL logic are examples of the potential in this area. Within the newer structures, it is becoming increasingly unrealistic to describe their properties in terms of equivalent transistor circuits, and more necessary to describe them in terms of their overall transfer functions, which might model a logical function like the storage cell. To achieve useful improvements it would seem necessary to move to semiconductor processes that require fewer masking steps and operations, since this both reduces processing costs and increases yield. The most promising techniques are those which exploit the possible three-dimensional structure of a semiconductor, with the tight tolerance features in the third dimension. An example is the VMOS process in which the channel length is defined by an epitaxial step rather than by a surface feature.

The improvements in circuit and process technology may be seen as complementing improvements in definition rather than as providing another route to increased circuit density.

### Constant-cost devices

The current pattern of advance is predicated on the existence of a high-volume demand for increased functional capability within a circuit. At present, circuit capabilities are inadequate to support a fully functional microcomputer, but this could be achieved by 1980. Such a device would satisfy the vast majority of the markets foreseen by the semiconductor companies, for example in the control of consumer products like cars. Thus, further increases in complexity are unlikely to generate the ever-increasing markets that the increased capital cost of higher-definition process lines will demand.

The exception to this conclusion is the use of microelectronics for the storage of the information. The current market for semiconductor storage in computers is clearly limited, both by the market for computers and by the architecture of mainframe computers, so that in the short to medium term similar arguments apply. In the long term, however, large new markets appear possible for the electronic storage of textual information within the office and the home, as a replacement for paper (see pp. 142-53) and for the storage of audio information (see p. 141). Thus in this area there could be the incentive to pursue the development of higher levels of integration.

In general, however, it is thought that the current pattern of development at constant prices will be replaced by two divergent approaches — the pursuit of very high performance at greatly increased cost, and the pursuit of low-cost devices.

## High-performance devices

There is a continuing demand for high-performance circuits. In terms of quantity, the demand is, and will probably remain, low. The main uses of such high-performance circuits are in highly responsive control systems (for example in military avionics or fusion reaction) and for high-performance computing. Substantial performance gains can be achieved by reduction in circuit size and by further increase in circuit complexity. To a first order, circuit delays are proportional to the size of a component, so that smaller circuits operate faster; also the use of more complex circuits can eliminate inter-package delays. The resultant circuits will present severe problems in terms of heat removal leading to expensive packaging; indeed further performance gains can be achieved by operating such circuits near absolute temperatures. All these requirements are very different from the general high-volume market for microelectronics discussed in the rest of this study, and it is to be expected that they will be satisfied by specialist suppliers. The markets to be served would tolerate extremely high prices per circuit, so low volumes would not be an obstacle, and it is to this market that electron-beam-machined circuits may be expected to be addressed.

## Low-cost devices

Given an adequate level of circuit complexity, probably around the 64K dynamic RAM (Random-Access Storage) capacity, it is possible to build a complete single-chip microcomputer that should satisfy the majority of high-volume demand currently envisaged. Thereafter, there would seem to be more benefit to be gained from expanding these markets by price reduction rather than attempting to create new markets by increasing circuit capability. Such an approach will require a rather different orientation on the part of the semiconductor industry, but there does seem to be scope for useful innovation.

— Maintaining the circuit at constant complexity and reducing chip size will reduce chip cost proportionately and also reduce the dissipation problems, thus allowing simpler packaging.

— The present process technologies are operated to produce a significant level of faulty circuits. The level of faulty circuits delivered to the customer is usually maintained around the 1% level by AQL techniques in the factory. This approach causes both manufacturer and user to incur costs. The manufacturer incurs costs because a significant proportion of faulty circuits are not identified before packaging, which is the most expensive part of the process, and the user incurs costs because it is necessary to test all the circuits before use. This requirement to test circuits is very onerous in a high-volume business like cars or toys, and the cost of the test equipment is escalating with the complexity of the circuits. It is to be expected that, in future, improvements in processing technology will be exploited to obtain increased yields, to an extent where AQL levels of .01% or better are achieved, thus obviating the need for the user to test circuits before assembly.

— Packaging costs currently dominate semiconductor manufacturing. This is because of the large number of pinouts (40 or more) used in most large-scale integrated circuits. Each pinout is expensive because of the manual effort required to make individual connections between the pinout and the bonding pad on the circuit, because of the high failure rate in this interconnection process, and because of the bonding area and drive circuitry required on the circuit itself. There is considerable room for reduction in cost by reducing pinout count and by adopting packaging techniques which eliminate manual bonding

52

in favour of direct automatic connection of the pinout on to the chip.

Using these kinds of approach, the semiconductor industry might be expected to deliver microcomputers in volume at a price of $1 in 1980 and 10c-20c in 1985.

## Exotic technology

Although nMOS technology would appear to represent the mainstream technology into the long term, a number of other technologies appear possible. Some of the important directions for research are:

— the use of cryogenics to improve performance and density
— the use of alternatives to silicon as a semiconductor (possibilities are indium phosphide or gallium arsenide)
— the use of novel electrical characteristics, like the Josephson device, in which a tunnelling junction is switched between normal conduction and superconduction by a magnetic current.

There is a wide variety of semiconductor phenomena which have been discovered but not explored in depth, and it is reasonable to expect that there is room for further revolutions in the technology. At present, the semiconductor phenomena used depend on bulk characteristics of electrons, such as current or voltage. The possibility remains to create devices which exploit the wave characteristics of electrons.

The semiconductor technology already available is more than adequate to satisfy the wide variety of applications discussed in this study, and its exploitation will lead to dramatic social and economic changes. It is not clear that there will be any immediate economic motivation to exploit more powerful characteristics of semiconductors.

As we have already said, electronics stands in the same relation to information as electricity does to power: it is a convenient, but not a unique representation. Although other representations, such as optical, are possible for information, the momentum associated with electronics, and the further developments that are possible, suggest that electronics will remain the preferred implementation into the very long term.

## Conclusions

*While there is technological opportunity for increased circuit complexity, the increasing capital costs and lack of obvious demand may cause the technology to evolve more slowly than in the past, with more emphasis on cost reduction than on increased circuit complexity.*

*There is still considerable opportunity for circuit and process innovation leading to lower costs and higher-density circuits.*

*If there is a divergence between low-cost and high-performance circuitry, then access to high-performance circuits may become difficult unless the U.K. develops its own capability (see p. 121-2). Such circuits will be essential for large computers and military applications.*


## STORAGE TECHNOLOGY

The continued development of semiconductor storage will enable it to replace low-cost magnetic media. This will be advantageous, because it eliminates one source of unreliability and because the access mechanism can be cheap enough for reasonable quantities of store to be associated with low-cost devices like digital telephones.

Economies of scale are expected to continue in storage technology, leading to the use of a variety of technologies and to cost advantages for bulk information storage. The divergence between active storage within a computer and archival storage that can be accessed by computer is likely to increase, leading to large storage systems being regarded as independent products.

### Hierarchy

There are clear economies of scale associated with current storage technology. As a result, computer storage systems are characterised by a hierarchical structure, and the use of a variety of technologies. The reasons for these economies of scale appear inherent. High-volume storage requires very high-density and near-zero energy retention, characteristics which are incompatible with the requirement for fast access.

At present, the objective is to design a storage hierarchy to create the image of a single-level storage system. In practice this is not achieved and storage systems have an inherent two-level structure based on the concept of the word and the file.

Rather than there being a convergence to a uniform store hierarchy, it is expected that the two-level structure will emerge as even more basic, with a clear distinction between active storage and archival storage systems.

## Active storage

At present there are four contending technologies — bipolar, MOS, CCD and bubbles. The first three are based on the electrical properties of silicon semiconductors, the last on the magnetic properties of garnet. The various technologies offer slightly different characteristics, so they are not in total competition.

The bipolar technology would seem to offer little advantage except performance. Since storage can be intrinsically slower than processing logic, there would seem to be no important large-scale use for bipolar technology in the longer term.

The other three technologies all have a comparable basic cell size and their fabrication is based on similar optical processing techniques. The primary difference lies in the overheads associated with access to the cell. These overheads dominate with small storage devices (e.g. 4K bits) but become less important as the size of the storage device increases.

In the medium term the bubble memory is seen to be at an increasing disadvantage, because of the very much higher cost of packaging. To operate, the bubble chip must be surrounded by a controlled magnetic field which requires separate coils and a permanent magnet. For this reason, it is not expected to become a dominant technology.

MOS and CCD are largely compatible technologies and can be mixed within the same device. As a result, they should be seen as complementary rather than as competitive, the circuit designer being able to trade off the properties to achieve appropriate external characteristics. Thus MOS technology is seen as remaining the dominant active storage technology in at least the short and medium term:

> — It is compatible with the circuitry required for digital television, which will be a large-scale market for storage technology.

- It is compatible with Read-Only Memory (ROM) technology which is seen as a large-scale future market.
- It uses the dominant semiconductor technology.
- It is compatible with logic circuitry facilitating the development of intelligent storage circuits.

In many ways the type of technology is less important than the size and cost characteristics of future active storage devices. Present state of the art is the 16K bit circuit with an access time of 250 nsec, a component cost of 0.03c per bit and a system cost of 0.1c per bit. Because of the regularity of a storage array, cell design can be highly optimised and greater component densities can be achieved. The storage device, is, however, subject to the same basic limitations as other semiconductor devices, leading to a short-term capacity limitation at 64 or 128K bits.

At present most storage devices present a very primitive one-bit interface. As the storage capacity increases, it becomes increasingly cost effective to incorporate more on-chip logic, leading to a more powerful interface. Indeed, at 64K bits or above, the marginal cost of incorporating a complete processor (equivalent to 4 or 8K bits in terms of chip area) becomes negligible. It is to be expected, therefore, that the current range of primitive storage circuits will be replaced by intelligent storage circuits. Whether such circuits will merge with the concept of a microcomputer, or remain a distinct species is not clear at this time.

An intelligent storage circuit might be expected to offer some, or all, of the following facilities:
- byte-oriented rather than bit-oriented output
- internal refresh logic
- serial input and output for data and address
- multiported operation
- error detection and correction
- automatic testing
- cache storage for higher speed
- access by name rather than address (i.e. implementing internal storage-management facilities)
- power minimisation by switching power only to the area addressed, or to areas that need memory refresh.

If we take the intelligent-store approach, much larger circuits could be fabricated with a consequent reduction in cost and improvement in reliability. To do this, discretionary addressing would be required to enable avoidance of faulty areas of the circuit. With this approach, it is possible to envisage circuits several inches square containing a few million bits as practicable in the mid 1980s, leading to system costs of 0.001c per bit.

Attention is usually focused on the cost per bit of storage. The minimum cost of a complete system is a factor of at least equal importance. The availability of low-cost storage systems containing 8K bits or more will have a revolutionary effect on the types of products that can be offered.

## Fixed storage

Read-Only Memory (ROM) is at present seen as a low-cost form of storage for microprocessors. The use of an aluminium mask to define information storage has enabled a higher bit density than is possible with Random Access Storage (RAM). The development of the discretionary-addressing technique will improve the competitive position of RAM, since it cannot be applied to the mask ROM storage.

There is, however, a more fundamental difference between the two types of storage. Information must be loaded into the RAM system bit by bit, whereas all the bits can be defined simultaneously with the mask ROM. This is just the difference between an audio cassette and a long-playing record. Magnetic tape is intrinsically the cheaper medium, but cassettes are expensive in comparison with records because they tie up expensive capital equipment while the information is loaded. Thus mask ROM is, and will continue to be, an appropriate technology for disseminating information. A book contains about 4M bits, an LP record contains about 32M bits. Semiconductor storage should become competitive with these technologies in the long term.

## Archival storage

At present, archival storage is based on magnetic media — disc, tape and, for low-cost applications, floppy disc or cassette. Various new contending technologies have been proposed, the most important of which are Electron Beam Addressed Memories (EBAM) and various types of optical systems:

    — In an EBAM memory, the semiconductor wafer is used as

a target for an indexed electron beam in an evacuated tube. It is thus an updated form of the Williams' tube, providing much greater information density, with present-day production prototypes offering a capacity of 32M bits. As a storage system it is inherently bulky and costly, and it does not appear appropriate as an archival system for the long term. There does seem some prospect that developments of this technology could provide for the very high-performance storage systems that will be required to complement a high-performance processor, since access in the third dimension could eliminate the stray capacity which slows down the operation of a conventional semi-conductor store.

— in a holographic store, information is stored on an array which can be indexed mechanically to a selected area and then be written or read using a laser beam. A holographic image is used because the system can then tolerate microscopic imperfections in the medium. This type of store has a number of advantages. As is not the case with the EBAM store, the medium can be replaceable since a vacuum is not required. Also, it has the potential to store or access a large number of bits in parallel, offering the possibility of high data rate. The difficulty has been to find a photochromic medium with adequate erasure properties to allow random access storage.

— If the requirement for erasure is removed, so that the storage is not reusable (like paper) then the holographic store can be used with a conventional photographic medium. Another alternative under development is the direct use of a laser beam to write on aluminised tape by vapourising the metal. This will provide a very high-density replacement for magnetic tape.

The cost of magnetic archival media is already competitive with paper or microfilm. That electronic information storage is not already in wide use is due to the following factors:

— The high cost of capturing information in an electronic form. The cost is high because the information is usually not available in electronic form and must be transcribed, which generates an additional cost. As electronic infor-

mation systems become more prevalent (see Chapter 6) this barrier will disappear, leading to a rapid increase in the demand for electronic information storage.
— The capital cost of storage systems can be extremely high. It is anticipated that the primary direction of development will be towards the provision of high-volume storage at lower capital costs, rather than towards the further reduction in media costs. On this basis it is the simpler storage systems that will be preferred.

Present electronic storage systems are at a severe disadvantage compared with conventional alternatives, since the medium is erasable. The use of erasable media is an historic consequence of the development of computers, and such media are probably inherently more expensive than fixed media. As an increasing amount of general information is held in electronic storage systems, so the demand will develop for such systems to hold information securely on a permanent basis, for example for legal or taxation purposes. Thus, it will either be necessary to provide secure methods to inhibit the erasure of reversible media, or to adopt nonreversible media. The latter course is obviously the better and it is thought that, in the long term, archival storage will be based on the use of permanent media; this makes the direct-laser aluminised-mylar store a very attractive proposition.

At the low end of the range for archival storage (for example, personal storage) the use of mechanical storage devices is undesirable, since there are inherent capital costs and since the reliability of such devices must be suspect. For these reasons, it is expected that low-volume archival storage will be based on some form of solid-state cassette. This is a potential market for the bubble store, but it will have to compete against semiconductor storage with some minimal energy source.


**Storage systems**
At present, storage is regarded as an integral and subsidiary part of a computer. As the volume of storage requirements increases and the cost of processing falls, the cost of large-scale storage systems will come to dominate. Also, the increased dependence of the community on electronically stored information will cause

greater emphasis to be placed on the long-term security of the information and on higher levels of availability.

For these reasons, storage will increasingly be regarded as an independent system providing specified levels of access, reliability, security and cost. To achieve these objectives, the storage system will be sold as a complete product, containing within itself a hierarchy of storage devices. The system will be controlled by processors dedicated to optimising the allocation of the information within the hierarchy and to providing archival and testing facilities. Such systems could be sold as independent products, and may well be marketed by companies other than the conventional mainframe-computer manufacturers.

The economies of scale associated with storage, and the need for long-term security for information could well lead to the development of data banks, in the properly accepted use of that term, that is, independent organisations which would, for a suitable charge, accept and store information, and indemnify the owner against loss.

## Conclusions

*Continuing cost reductions of an order of magnitude or more are possible within the various storage technologies. The reduction in cost per bit may be less important than the reduction in the minimum cost for a particular type of storage.*

*For high-volume archival storage, the highest densities will be required and this means that tape in some form — magnetic, photographic or optical — will be the preferred medium.*

*There is still opportunity for innovation, particularly in the area of bulk storage systems.*

*There will be an emergent market for storage systems as products in their own right.*

## PERIPHERALS

The term 'peripheral' is somewhat ambiguous, the connotation being 'that which is attached to a central processor'. For the purpose of this discussion, the term 'peripheral' will be restricted to mean a transducer — that is, something which transforms the medium of representation of information.

60

## Terminals

The rapid growth of visual terminals is reducing the demand for more traditional methods of input and output. In the long to very long term, the availability of an extensive communications network and the universal use of terminals could eliminate the use of paper as a medium for information representation, thus eliminating the need for many of the types of peripherals in use today.

The change from a paper-based to an electronic-based information economy may be more rapid than expected, because the interim period with dual systems involves unnecessary transcription costs. As discussed in Chapter 6, the technology should rapidly reach the stage where low-cost electronic terminals are available, and the primary limitation may be in the provision of adequate communication capabilities.

It is obvious that the processing power available within the terminal will increase as terminals are based on more powerful microcomputers. This power could be used in a variety of ways:

— to perform a larger part of the work required by the user, reducing the dependence on remote computing power
— to improve the interface between the user and the system, using the processing power to simplify the interactions for the user and making the system more tolerant of user's errors or idiosyncracies
— to improve the interface between the terminal and the system, and make the terminals capable of supporting more powerful protocols
— to reduce the cost of the terminal by simplifying the display and interface electronics and by reducing the mechanical requirements on the keyboard.

It is to be expected that a terminal will incorporate several microcomputers, each attending to a separate function — keyboard, display, computer interface, human interaction and task execution.

The main limitation on the cost of a terminal is the need to use a cathode ray tube (CRT) display. This adds to the cost because of its power requirement, its bulk and the associated electronics, apart from the cost of the tube itself. In the short term, it is to be expected that CRT displays will be improved to provide a complete A4 page of information. This is just feasible using a standard television monitor and a $5 \times 9$ resolution dot-matrix display.

One of the prime areas of concern must be whether present CRT technology provides an environment which is acceptable for continuous use by operators. There are some obvious disadvantages:

— The display is luminous, which raises difficulties with lighting and reflections.
— The picture is not completely stable.
— The poor quality of the characters makes reading difficult.

Such disadvantages were acceptable when the terminal was only used intermittently by highly motivated staff. They may well not be acceptable for a device intended as the primary working tool for many people, which will be operated continuously throughout the day.

There are thus many reasons for wanting to replace the CRT by some more amenable display technique. The most desirable contender would be some form of solid-state display, preferably based on absorption rather than transmission characteristics. Various forms of flat display have been under development for television for some time, and recently Hitachi demonstrated a black and white television based on a small $5 \times 3$ inch display. One of the main difficulties with such displays is to reduce the number of external connections to an acceptable level. In principle, the requirements for a visual display are less demanding than those of television, and it is to be expected that flat displays will be developed for terminals before they find applications in television.

Ideally, a terminal for human use should match the primary communication characteristics of man. Present terminals achieve this for input to the human, where the primary channel is visual, but not for output, where the primary channel is speech. For reasons discussed on p. 176-8, it is not expected that general speech input will be feasible, at least before the long term. Because of its cost and band-width advantage, it is expected that the keyboard will remain the primary method of input for both the professional and the lay user.


**Printing**
With the range of uses anticipated in Chapter 6, it is expected

that the demand for printed output will increase in the short and medium term. The primary demand will be for:

— low-cost printers to provide office-quality printing for letters and other items, printed one-off or in low volume. The most applicable mechanical technologies would seem to be the daisy-wheel printer and its variants, since the dot-matrix and related printers provide inadequate quality. The alternative is the development of xerographic output. Since direct xerographic printing is also required within the office during the interim period until paper is eliminated, it is expected that office-copier technology will be developed so that the same copier can be used in a multi-purpose role:
   — direct copying of originals
   — facsimile transcription
   — character-coded transmission
   — direct computer output
— high-volume printers to provide multiple outputs to replace conventional printing in low- and medium-volume situations. Here, the ink-jet technology offers a trade-off between speed and quality and would make colour printing possible. The alternative, as above, is the use of xerographic principles.

Here again, the availability of the microcomputer may be expected to have an impact on printer design, both by simplifying and improving the mechanical design and by associating more facilities with the printer itself — for example, enabling it to format its own output, and improving its interface with the system.


**Transducers**
One of the primary problems in the area of microcomputers is to provide transducers to enable the microcomputer to collect or transmit physical information at an acceptable cost. At present, the majority of transducers in general use convert between the physical measurement and an analog electrical voltage. As a result, a further analog-to-digital conversion stage is necessary, which adds to the cost. To combat this, the conversion process will migrate on chip as far as possible. In the case of integrated injection logic it is possible to achieve complete

analog-digital converter on-chip, allowing input of analog voltages. Here nMOS logic is at a disadvantage, since it can only handle the logical aspects of conversion on-chip, and must rely on an external voltage comparator.

In the long term it may be expected that there will be development of direct digital transducers, for example relying on frequency measurements. An example of such development is the recently discovered Wiegand effect.

### Interfacing

During the earlier stages of computing, the peripherals were an integral part of the computer. As computing developed and there was a need to vary the configuration to satisfy different user requirements, so the peripheral became an independently connected item, initially through dedicated ports and subsequently through a general system interface.

Since that time, there have been three lines of development:
- increasing de facto standardisation of the low-level characteristics of each peripheral, enabling different devices (e.g., floppy disc mechanisms) to be used interchangeably and thus providing a second-source option to the manufacturers of computers.
- the growth of the plug-compatible market, enabling end users to buy compatible peripherals for de facto standard computers like the IBM 370 and the Digital PDP 11 from independent suppliers.
- the use of the telecommunications standard V24/RS232, to provide a general standard for the interconnection of low-speed peripherals both locally and remotely.

In the future, it is expected that the use of communication standards as the basis for the interconnection of peripherals will become dominant, replacing manufacturers' standard interfaces and standardisation at the mechanical level. It is also expected that the standards will move away from the low-level V24 standard to high-level communication standards and protocols, probably based on X25 and its associated standards.

### Conclusions

*The terminal is expected to become the dominant means of*

*communication between man and system. Its widespread use may be limited more by the availability of communication facilities than by cost.*

*There must be questions whether or not current CRT terminals provide a satisfactory ergonomic and physical environment for continuous use.*

*There is a continuing opportunity for new companies and new products in the peripherals area.*

*In the long term, high-level communication interfaces will be the primary method for accessing peripherals.*

## COMMUNICATIONS

Communications are an essential aspect of information technology. At present we lack the knowledge to design a cost-effective data communications system suitable for the very long term, and there is insufficient research and development in this area.

### Economies of scale

Communications systems are characterised by marked economies of scale. High-bandwidth transmission systems offer much lower cost per bit than low-bandwidth systems. With recent technological developments, for example in optical transmission, these economies of scale have become more marked. As a result, large communication systems are more economic than small systems, and communications systems are organised in a hierarchy of levels to concentrate the information flow over a few very high-bandwidth lines. A further result is that the cost per bit for stations with low-bandwidth requirements is always greater than the cost per bit for stations with high-bandwidth requirements, because the majority of the cost is incurred in the local interconnection.

Thus, although telecommunication technology is advancing rapidly, this is unlikely to be reflected in much reduced costs for low-bandwidth users. Indeed, taking into account the high costs of installation and maintenance, it must be expected that telecommunication charges will tend to increase in the future.

### System organisation

The already high cost of telecommunications is a major factor in the design of distributed computer systems. The reducing costs of processing and storage contrast with the static or rising costs of telecommunications and will make the economics of such systems even less attractive in the future. Ignoring the details of tariffing, it would seem that communication is inherently expensive in comparison with processing. For one computer to telecommunicate with another, the information must pass through a series of four or five telecommunication switches, each of which is likely to be of the same order of complexity as the terminal computers themselves. Thus, even excluding line costs, the depreciation charge on a telecommunication system is likely to exceed the depreciation charge on the local equipment, perhaps by an order of magnitude.

The economies of using telecommunications access will depend on the frequency with which a particular facility (e.g. program or piece of data) is used. Only if the frequency of use is low will remote access be economic in comparison with provision of the same facility locally.

This situation contrasts strongly with the pattern of telecommunication usage to data. The main data traffic has been generated by access to remote-processing facilities, particularly through remote batch terminals for interactive time-sharing terminals. This type of usage must be expected to decline in favour of the use of data communication for:

— access to remote services which it is uneconomic to provide locally
— transfer of information

This change in usage has strong implications both for the organisation of computing facilities and for characteristics of traffic on the data-communication network, and hence on its design.

## Packet switching
Most data-communication systems under development are based on the concept of packet switching. The rationale for packet switching is that the length of a message is very short in relation to the time necessary to establish a conventional end-to-end call, and that it is better economics, therefore, to transmit that message stage by stage through the network, storing the message at each node until the necessary line capacity is available for transmission

to the next node. Thus packet switching is the electronic analogue of a postal service, where letters are posted into the system and will be received after a variable delay, having been sorted and stored at various points along the route. For short messages, packet switching provides better utilisation of transmission, line and switching nodes than do conventional telecommunication switches; the disadvantage is that the complexity of the switch itself is greatly increased. This is particularly so if the requirement is placed on a network that it should deliver messages in the order in which they were transmitted, something the postal service does not aim to achieve.

International protocols for packet switching (X25) have been agreed by the Consultative Committee for International Telegraph and Telecommunications (CCITT), and further standards — for example on virtual terminals — are under discussion.

Although there has been considerable pressure from the computer community for the PTTs to provide data communication using packet switching, and although packet-switching systems are now being installed in the majority of the major industrial countries, there have been no good analyses of the economics of such networks in comparison with alternatives. Equally there have been no proper studies done on the efficiency of the X25 protocols, or indeed on their validity or limitations. All the thinking about packet-switch networks has been based on the comparatively small-scale networks that are appropriate to the 1970s and 1980s. The characteristics of a network change rapidly with the scale of that network, and the current approach may not be appropriate for a national network handling a similar volume of activity to the present voice network. It is extremely alarming that the data-communications network of the future could be prejudiced by the lack of proper study and evaluation at this stage; there is a need for a properly coordinated programme to study the most economic arrangements for future data networks. In a very large-scale network, it is the cost and utilisation of the local interconnections which dominate. Because of the economies of scale, the trunk distribution becomes a negligible part of the cost. This is illustrated in the voice network, where the local connections represent 85% of the capital cost at the margin and yet each connection has an average utilisation of less than 0.1%.

**Local networks**

The primary problem, as indicated above, is to provide cost-effective local-distribution networks, both from the local exchanges to the subscriber, and within the subscriber's environment. For the present, only offices and factories use private switches, but as the range of communicating devices is extended, it is reasonable to expect that each home will require its own local distribution, providing connection between rooms and the connection of multiple devices, such as the conventional telephone, electronic typewriter (see pp. 145-6), electric meter and gas meter, to the system.

The obvious solution is to use a high-speed serial-transmission line, with active sockets, each containing a semiconductor switch to enable information to be multiplexed onto or off the line. Thus the data-communication system of the future may look far more like the electricity-supply system in the home than like the current telephone. For most purposes there would seem to be no need to go to optical techniques, since conventional coaxial or balanced pairlines would provide adequate bandwidth. One attractive solution would be to use a radiating cable and RF techniques to eliminate the need for physical interconnection between the information devices and the information systems.

This general area of multiplexed high-speed interconnection between many devices is a key to the provision of economic data-transmission systems. It requires far more effort than is currently being directed towards solving the problem.

**Standardisation**

The X25 family of standard protocols is obviously going to exert great influence over the development of computing. The provision of standards, initially for a virtual terminal, and possibly for job-control language, virtual files and virtual processes, must in turn be expected to lead to far more standard external computer characteristics. This is in no way undesirable, but it must be recognised that the definition of computer characteristics through X25 is likely to influence real computer characteristics, since in order to support X25 it will be necessary for operating systems to support both virtual and actual facilities, and in this case it would obviously be desirable to make the actual facilities coincide with the virtual. For this reason, all countries should be taking an active interest in the definition of these standards, to ensure that they will lead to efficient usage of computers in the long term.

## Conclusions

*Too much effort in communications research and development has been directed towards the problem of high-speed transmission, rather than to achieving good economics for the transmission of low-speed data.*

*Work is still necessary to determine the most appropriate form for large-scale data networks, both in terms of multiplexing techniques and in terms of protocols.*

*Research and development are required to provide economic forms of local data distribution.*

*Research and evaluation is still required on the X25 family of information protocols.*

## MICROCOMPUTERS

The microcomputer is seen as the key technological development, because it enables information products to be designed at the program level rather than the logic level. It is conjectured that the momentum to *de facto* standardisation may prevent any substantial advance over existing microprocessor design.

### Status quo

The concept of the microcomputer — a computer on a chip — is almost as old as the idea of the integrated circuit itself. In the mid 1960s many forecasts were made of the development during the 1980s of computers on a chip.

Set against this background, the 'invention' of the microprocessor by Hoff in 1971 at Intel should not have been surprising — after all, Robert Noyce, one of the founders of Intel, had been amongst those predicting the development of a computer on a chip. And yet, no other company attempted to develop the microprocessor, and the proposal from Hoff was a surprise to his own management. The answer would seem to be that the semiconductor industry had not appreciated that microprocessors would find a different class of application to conventional computers, and that even a very primitive processor could have many large-volume uses.

Since the development of the Intel 4004, a large number of manufacturers have entered the microprocessor market, so that there are now over fifty distinct microprocessor architectures available, while the number of separate offerings totals well over 100. Only a few of these are sold in volume, and the market is exhibiting a strong tendency to *de facto* standardisation (a subject discussed further in Chapter 5).

Alongside this increased variety of microprocessors, the complexity and capability of microprocessor designs has increased with the advance of semiconductor technology. In transistor count, the complexity of a microprocessor is considerably below that of a comparable storage circuit. There are two reasons:

— The microprocessor requires 'random' logic which achieves a much lower density than a regular array of storage cells, perhaps by a factor of five.
— The design and development of a microprocessor is considerably more difficult than that of a storage device, so that microprocessors take longer to develop and use older technology.

This increase in complexity has been used in various ways:

— to increase the word length of the microprocessor (from 4 to 8 and sometimes to 16 bits)
— to increase the variety of the instruction set (a somewhat questionable improvement)
— to increase the functions on the chip, and so reduce the amount of external circuitry required to make the device useful.

It is this last development which is the most important. The earliest microprocessors required a large amount of external circuitry to turn them into usable devices. Typically, each microprocessor required 30-50 additional circuits before it was viable. The number of additional circuits required has been reduced, firstly by bringing all the processor functions on chip, like interface controls and clock generation, and secondly by bringing other functions on chip, like storage and interfacing. The advantages of this approach are that:

— It reduces cost by reducing the number of components.

- It reduces cost by reducing the associated packaging costs for the circuitry
- It reduces the expertise necessary to design a product based on a microprocessor.

The combination of these factors has meant that the cost of microprocessor-based systems has fallen more rapidly than is indicated by an examination of the component prices themselves.

In 1977-78 microprocessors were available in three basic varieties:

- the microprocessor, requiring external store and interfacing. Examples are the Intel 8080, Motorol 6800 and Texas 9900.
- the combination circuit, in which the microprocessor is combined with storage or interfacing, thereby reducing the overall package count in a system. Examples are the Intel 8085, the Motorol 6802 and the Fairchild F-8.
- the microcomputer, in which all the functions of the computer processor, program store, data store and interfacing are combined on a single circuit. Examples are the Intel 8048, the Texas 9940 and the Mostek 3870.

With the microcomputer, the semiconductor industry has truly achieved the goal of a computer on a chip, although with the present scale of integration, the processors offered are extremely limited in capability.

## Development of the microprocessor

The first microprocessor used a 4-bit word. As the technology has advanced, 8- and 16-bit devices have been introduced. Two alternative views have been put forward for the pattern of further development. The first is that each type of microprocessor has a natural market and will continue into the future:

- 4 bits for control systems
- 8 bits for information products
- 16 bits as a minicomputer replacement.

The alternative view sees word length as a limitation, and looks forward to the successive replacement of 4-bit devices by 8-bit, 8 by 16, and, in the future 16 by 32 or 64.

In the short term, most semiconductor companies propose to introduce 16-bit products, which must necessarily be incompatible with their existing 8-bit products. The rationale behind this move seems to be based largely on competitive pressure — the technology allows this capability, and manufacturers feel that if they do not introduce leading-edge products they will lose the future market. As yet, it would seem that few of the semiconductor manufacturers appreciate the high cost associated with existing architectures, and they are still willing to write off early designs. Several such new products were announced in 1978, with deliveries starting in 1979 and volume in 1980.

As time progresses and cost of supporting a microprocessor architecture is becoming more apparent to the manufacturers, their attitudes are gradually changing. One company, Texas, already has a compatible range based on a 16-bit architecture, and it may have established a commanding lead in the market. It seems possible that the other manufacturers may eventually decide to de-commit themselves from the 16-bit market and remain with their existing product lines, using technology to enhance their capability in other ways.

The semiconductor developments that are envisaged will make possible the complete 16-bit microcomputer with a 4K byte store by 1980. Such a device would have a performance in excess of most present-day minicomputers. It is argued in Chapter 5 that there are strong market forces that will lead to a reduction in the number of microcomputer architectures, leaving just one or two *de facto* standards, just as has happened with the computer (IBM 360) and the minicomputer (Digital PDP11). The existence of a standard architecture does not preclude its being offered in a variety of forms with different capabilities and cost, and it is expected that one or more microcomputer families will emerge as the dominant product, offered by several, not necessarily compatible, second sources.

Because of the high cost of developing new architectures, semiconductor manufacturers will be increasingly attracted to the idea of copying existing computer architectures, thereby obviating the need to develop much of the support software. To date, this approach has not been practicable, because the complexity of even a simple minicomputer has been rather greater than could be achieved by the technology. This situation is now changing. In order of complexity, the three obvious architectures

72

to copy are:
- the Data General Nova
- the Digital PDP11
- the IBM 360.

Neither the software, nor the organisation of these computers is suited to microprocessor technology or application, but the advantages of a wide range of software and expertise is likely to overwhelm these drawbacks.

Fairchild has already announced a copy of the Nova, and Data General has initiated litigation to prevent the use of its architecture. Unfortunately, this is an extremely ill defined area of the law, and it is not clear whether, or to what extent, a company can protect an architecture or the associated software. For the present, the threat of litigation has been sufficient to deter most would-be copiers, but the pressure to copy is likely to increase. This is particularly so since the PDP11 architecture has been adopted as a standard by the American Department of Defense and copying the IBM 360 architecture at the discrete-logic level is already an accepted practice.

As a guide, the technology has the capability to integrate existing computer architectures fully at the following dates:
- 1978 Data General Nova
- 1979 Digital PDP11
- 1980 IBM 360.

The microprocessor versions of these architectures are likely to offer higher throughput than the basic discrete versions, although they will not match the throughput of the higher members of these computer ranges.

### The microcomputer as a universal component

The importance of the microcomputer is that it provides a new level of abstraction in the physical design of information systems. So far there have been two levels of abstraction:
- the electronic component: here, information is represented by an electrical signal and the design is carried out in terms of electrical properties
- the logical gate: here, information is represented by logical levels, and design is in terms of a logical calculus; the electrical details have largely been abstracted from

the design process, although not entirely, because of the imperfections in the logical components.

The microcomputer offers the potential for the design of information systems in terms of a third level of abstraction, based on language — where the basic unit is the word, which can be given specific semantic connotations by the provision of an appropriate set of information operations. As yet, the microcomputer falls short of being an adequate information component for two reasons:

- It is a very imperfect abstraction from the logical and electrical levels, with the result that its use requires detailed consideration of logical and electrical properties as well as its information properties. The solution to this difficulty is to formalise the communication between the microcomputer and its environment, so that it operates correctly at the information level, communicating words of information. This means that communication with a microcomputer should only occur through a rigorously defined interface.

- Whereas at the logical level there is a ready-made mathematical calculus for design, there is no such ready-made calculus at the microcomputer level. Such a calculus would provide the formalism for representing the interaction between a number of intercommunicating devices which operate autonomously. Until such a calculus is developed, the design objectives of microcomputers will remain intuitive rather than scientific. Such a calculus is likely to take the form of a programming language that has primitive operations making possible communication between parallel processes. The primitive elements of such a language may be seen as akin to the axioms of a mathematical theory, enabling a much wider variety of user languages to be constructed from the primitive elements.

The significance of the microcomputer is not widely appreciated. This is partly because the design technique for using this component has not been developed; it is also because the emergence of the microcomputer has been obscured by the intermediate version of the microprocessor which has been available with increasing capability since the early 1970s.

## The future of the microcomputer

The semiconductor industry is achieving an increasing capability to integrate circuitry. Potentially, the circuitry could be organised in a wide variety of ways. Will the microcomputer represent a large proportion of semiconductor usage, or a diminishing proportion if more specialised circuits are developed for specific applications?

It is often argued that, as markets become better defined, products initially developed using a microprocessor will be replaced by custom circuits with random logic designed specifically for the purpose. In practice, the trend has almost invariably been in the opposite direction. Examples are the calculator (which was initially a custom circuit), car ignition and, more recently, the watch. In each case, the original reason for using custom logic was that the programmed version would cost more. However, as the level of integration has increased, it has been found advantageous to move from the custom circuit to the microprocessor because the latter offers a much greater degree of flexibility and a wider range of functions for the user.

It is expected that this trend will continue, and that most information products will be based on programmed microcomputers. There are various reasons for this:

— In terms of semiconductor area, the programmed microcomputer is very efficient compared to custom logic. This is because a large amount of the information is stored in a compact regular array, rather than in the irregular low-density patterns characteristic of random logic. Thus the advantage of custom logic over microcomputers is likely to lie in terms of performance rather than cost, and in most cases performance is not essential.

— The development of the programmed microcomputer is far easier, particularly where modifications are required (for example, because the specification has changed). This is true even now, when the development aids are relatively primitive. The advantage of the microcomputer is likely to increase in the future.

— The use of a standard microcomputer architecture greatly simplifies the problem of testing the finished device.

Indeed it will often be economic to incorporate test programs into the product itself.

It is not expected that custom circuits will disappear, but rather that their nature will change. Instead of implementing custom logic, or a special application-dependent microprocessor, such circuits will be based on the use of a standard architecture, and will provide a configuration specific to the application in terms of the quantity of storage used and the types of interface. In this way, the user of the custom circuit will be able to ensure that his component cost is minimised, while still being able to take advantage of the characteristics of a standard programmable architecture. Such custom circuits would be easy to develop, because they would consist of standard sub-masks interconnected, perhaps on a bus, to form the mask set for the device.

The volume necessary to justify customisation may be expected to increase, as the degree of production automation increases, and component costs fall. For the vast majority of applications — for example in process control — it may be economic to under-utilise a standard device, rather than to have a range of devices with varying capability.

The characteristics of the microcomputer will be optimised around the requirements of the high-volume market, and not around conventional computer usage. For this reason, micro-computers may be expected to diverge increasingly from the established concept of the computer, and may not be an ideal component for conventional computer design.

Four independent factors may operate to stabilise the development of the microcomputer in the medium term, and perhaps cause it to ossify in the long term:

— The simple microcomputer at the 1980 technology level will be adequate for the majority of high-volume applications that have been proposed.

— There is no obvious larger-scale building brick than the microcomputer. Elaboration of the concept of the micro-computer by providing more complex configurations is likely to prove self defeating, because the variety of such configurations is too large to be economic. (The same problem limited the exploitation of small- and medium-scale integration during the late 1960s and early 1970s.)

76

— As the microcomputer market matures, the introduction of new architectures will become progressively more difficult. Further change may be inhibited by the market mechanism of product dominance (see Chapter 5), even though technologically feasible, and even desirable.
— Semiconductor technology is reaching the stage where the return from further levels of integration may start to decrease.

It will require a major shake-up to overcome these stabilising factors. Although numerous technological advances can be foreseen, none of these is likely to be of a magnitude to justify its introduction, so that excluding any fundamental advances which make current computer concepts obsolete, further change may not be possible through normal market mechanisms and, if thought desirable, may require government intervention.

## Computer architectural developments

Present microprocessor designs are very unsatisfactory. They have, in general, been developed without taking cognisance of the design experience of computers and minicomputers, and as a result repeat many of the early mistakes. It should be recognised, however, that a microprocessor is not just a computer writ small, and that copying good computer practice will not necessarily give the optimum design for a microprocessor. There are various reasons for this; the following is a selection of the more important:

— In a conventional computer, the emphasis is on internal computation, rather than on external interaction. As the scale of the computer is reduced, first to a minicomputer then to the microcomputer, the importance of the internal operation declines in comparison with the importance of the interface operation. It is as though the computer could be represented by a circle, with the interior for internal processing and the circumference as the interface. As the size of the circle is reduced, so the properties of the boundary come to dominate. This consideration affects both the organisation of the internal architecture and the demand for programming languages.
— Measured in silicon area, the cost of processing is declining in relation to the cost of storage — not vice versa,

77

as is widely assumed in the computing industry. A simple processor equates to about 0.5K bytes of storage, a POP11 to 1K byte and an IBM 360 to 2K bytes. On this basis, it is not very good economics to associate a large storage system with a single processor. Better to find ways of using more processors within the system, even if only for trivial functions like address mapping. This size relationship appears intrinsic and in the long term it could lead to radical rethinking about the large and monolithic programs which dominate conventional computing today.

— There is an even less favourable relationship between procesing and interface capability, measured in silicon area. Because of the space consumed by the drive transistors and the bonding pads, 16 bits of interface occupies a similar area to a simple processor so that, again, there is a strong rationale for trading interface for processing wherever possible. This will lead to a predominance of serial interfacers and information-compression techniques.

— The cost of random logic is high in comparison with storage. As a result, the most economical approach to microprocessor design is to use a very simple logical organisation, backed if necessary by a microprogram level to provide the user-instruction set. For reasons explained later, it may be advantageous to give direct access to the simple instruction set rather than provide the user with the complete instruction set associated with minicomputers and computers.

— The range of applications is different. Conventional computers have been dominated by numerical calculation. In the majority of microprocessor applications, current and foreseen, numerical calculation is not a dominant factor. Most microprocessors are concerned with handling information, and within human organisations this is represented in alphabetic form.

**Word length.** The concept of word length is imprecise. Current 8-bit microprocessors like the Intel 8080 do provide some 16-bit operating capabilities, and 16-bit microprocessors like the Texas 9900 offer 8-bit operations. This fragmentation of the concept of

78

word length, with the microcomputer able to handle a number of units of different size, may be expected to increase. Thus it is more relevant to ask what range of information units the microprocessor computer should be able to handle.

Within a conventional design philosphy, 16 bits would appear adequate, since this provides sufficient address space for the majority of applications, and gives a good balance between the store and processor in terms of silicon area. 16 bits is not adequate for numerical calculation, but neither is 32 bits. A better alternative would be the use of multilength decimal representation for numbers, and this approach has already been used on some microcomputers.

With a more advanced organisation, the word length might be reduced to 8 bits or less, using hardware assistance for multilength operations on larger quantities.

**Addressing.** Mapping names into a linear address space is a primitive technique. It reduces the efficiency of representation and removes the contextual structure which is necessary for protection and for certain types of access. The direct support for structured names based on decimal or alphabet strings appears a logical development for microcomputer architecture.

This approach is particularly relevant when the area relationship between store and processing is considered. Rather than constructing stores with large address spaces, it would seem better to fragment the store between a number of processors, which can provide structured access to information at little additional cost.

One alternative often proposed is the use of content-addressable storage. This is not considered to be as attractive an approach because:

— The content-addressable logic is badly under-utilised
— It seems desirable to move away from the bit-level approach, to an information-oriented approach, where the microprocessor has direct understanding of the character set.

**Code compression.** Currently available microcomputer architectures are already fairly close to the limit of code compaction. It is probably not possible to reduce the size of code by more than another 30%, so that improved architectures will rapidly run into diminishing returns.

**Data compression.** The storage of data is extremely inefficient. The use of 8-bit characters to represent text is extravagant. Replacement of the 8-bit character by a 5- or 4-bit character, and the elimination of format in redundancy, could cut the size of text storage by a substantial factor. Already much information is stored electronically, and the volume will increase rapidly. It is to be expected that the storage of information will represent the major capital cost of information technology in the future, so the adoption of a more efficient representation for information would lead to large-scale savings in the longer term.

The handling of numerical information within computers is unsatisfactory. Numbers are normally mapped into binary fields with arbitrary constraints on size. A more appropriate method would be to use direct decimal representation, which can readily be implemented in a microcomputer.

A very attractive approach would be to set the primitive operation level of the microprocessor at the alphabetic level, so that it had a direct concept of an alphabet and of associated primitive operations like decimal addition, naming and function call. If this were done, the concept of binary representation and binary operation would be totally eliminated from the instruction level of the microcomputer.

**Direct execution.** The use of computers designed for direct execution of high level languages has often been proposed but has not been adopted in practice. The primary difficulty in the case of conventional computers has been the need to support a variety of languages on a single computer, which largely invalidates this approach. The microcomputer does not have the same need to support multiple languages, so the direct implementation of a high-level language is more attractive. Such a system might either execute a compiled form of the language, or an interpreted form. In practice, there is a continuum between these two extremes, and the low-cost logic would favour an approach biased towards the interpretative end, which in turn could affect the characteristics of the language used.

**Parallelism.** There is an inherent need for parallelism in a micro-computer, since it must support not only its internal operations but also operations at the interface. At present this is done by interrupts which timeshare the processor between internal and external operations. On some of the latest microcomputers separate register sets have been provided for internal and external operations, thereby reducing the overheads caused by interrupts. Given the improvements in processing and storage, it may be expected that future microcomputers will use separate programmable processors for internal and external operations. These processors would communicate through common storage using some semaphore mechanism. In practice there might be several external processors, one for each interface channel.

It is also possible to envisage a higher degree of internal parallelism, but the extent to which this is exploited will depend upon the development of appropriate languages and data-flow concepts.

**Interface.** The pinouts of a microcomputer may be regarded as forming an interface. In general, the definition of the operation of these pinouts is far too irregular and uncontrolled to be regarded as a satisfactory interface. The definition of a proper interface for the microcomputer would greatly facilitate its use singly and in assemblies, as well as reducing the pinout count. For most purposes, the microcomputer could use serial interfaces in place of parallel, and this would further reduce the pinout count. Pinouts represent a major cost in chip area and packaging so the development of improved microcomputer interfaces could lead to useful cost reductions.

**Support.** The level of technical support for the microcomputer is improving as the need for development aids is more widely appreciated and the cost of programming errors becomes apparent. For the present, the programming languages available are woefully inadequate. Existing computer languages concentrate on internal algorithmic structure rather than the control of external events.

The attempts by the semiconductor manufacturers to adapt existing computer languages have been disastrous. If, and when, good programming languages are developed for microcomputers, the trend towards the use of high-level languages may be expected to gather speed.

## Conclusions

*The microcomputer — a complete computer on a chip — is seen as a very significant development, both because it can cut the cost of many potential microprocessor applications, and because it represents a new level of abstraction for the design process.*

*In the medium term, the level of integration from the semiconductor industry will be adequate to achieve competent microcomputers, and there may be a trend for the semiconductor manufacturers to copy existing computer architectures like the POP11 or the IBM 360.*

*A variety of architectural developments are possible in the medium to long term, but may be inhibited by the pressure to de factor standardisation.*

## LANGUAGES

Language is the primary interface between the user and the computer, and between one computer and another. Good languages may be expected to make the use of information systems more efficient and, as such, are in the best interests of the user. The market mechanisms and standardisation methods that exist at present are unlikely to lead to improved languages.

The importance of programming languages is often misunderstood. Bad programming languages lead to bad programs and programmer inefficiency. Good programming languages do not necessarily lead to good programs or great efficiency. The introduction of high-level languages provided a great improvement in productivity, as compared to assembly language programming. There is little evidence to suggest that further developments of programming languages could improve programming performance more than marginally, given the current level of understanding of language design. From the point of view of productivity, effort could better be directed to other aspects of programming.

Languages are important because they represent the primary interface with the computer, and they must inevitably be the subject of long-term universal standardisation. Unless adequate standards are established in the medium term, the ultimate efficiency and convenience of computing may be prejudiced. There is an obvious analogy with natural languages, where *de facto* standardisation on a variety of incompatible languages has led to a great deal of unnecessay inefficiency and cost. A more alarming analogy can be drawn from the family of Indo-European languages, where it is possible to deduce the main features of the primal language from the features of its descendants. There is the risk that future computer linguists may be able to deduce the architectural form of the IBM 704 if the current pattern of language evolution continues.

Up to the present, computer languages have been concerned almost exclusively with programming. In some sense, this is only one aspect of languages; the emergence of data-description languages and specification languages indicates others. It is important that these various aspects be seen as part of a more generalised language system and that they are not developed or standardised independently.

### Language developments
It is useful to separate language features into two classes — deep features, which determine the formal characteristics of the language, and surface features, which largely determine the ergonomics of the language for the user. Discussions about the merits of various languages often founder on the failure to distinguish these two aspects, the language theoretician emphasising deep features while the user is concerned almost totally with the surface. This may be one reason why languages like Fortran, Basic and APL achieve a high degree of popularity while being condemned by the experts.

Understanding of deep features is growing slowly, but until a better understanding is obtained there is little hope of providing radically improved languages. At this level a number of developments should be expected in the medium to long term:

    — A better understanding of the sequencing of computations, leading to the adequate representation of concepts

83

like parallel computation, asynchronous interrupts and communication external to the program.

— A more rational partitioning of responsibilities between programming languages and operating systems. Many facilities appear within the operating system because it has not been possible so far to codify them to a level where they can be incorporated into languages. The trend towards implementing parts of existing operating systems into hardware as read-only memory will make progress in this area difficult.

— Modularisation. The provision of improved facilities for building programs from smaller units and for making changes to existing programs. This involves the introduction of new structural concepts to replace or extend concepts like block structure, procedures and processes.

— More appropriate primitive entities. Entities handled in present programming languages are abstractions of the features of current computer architectures (e.g., integers and floating point numbers) rather than a representation of more general concepts.

— Better methods for defining new entities within a program. Such entities might be program, or data, or a combination of the two. This should bring together the concepts of programming languages and data-description languages; it should also realise the concept of an extensible language enabling the same deep structural concepts to be used for primitive programming concepts and for application-oriented constructs. There would then be no need to regard application-oriented languages as being distinct from general programming languages.

— Improved translation techniques with languages defined within specific grammatical constraints. The ready availability of processing capability, and a higher degree of interaction, will encourage a trend from compilation to interpretation. This will cause a re-evaluation of the merits of binding various aspects of a language at compile time or run time, and will probably lead to greater flexibility in the time of binding.

No comparable sequence of improvements can be foreseen

for the surface features of language. Virtually no research or development is being undertaken in this area, even though it has more direct relevance to programmer productivity and language acceptability. It is noteworthy, for example, that highly structured languages, for all their advantages, appear ergonomically unsatisfactory when judged by the evident preference of users for unstructured languages like Fortran or APL. Perhaps such structuring, if necessary, should not be made directly apparent to the user.

Within these theoretical developments, a general trend should be expected towards the use of specification languages, interactive operation and semantic content within language.

In a specification language, the user defines the result that he wishes to obtain, rather than the method by which it is obtained. Thus, for a specification language, the expression $Y^2$: $= X$ may be acceptable, since given a value for Y, it is possible to verify that its square is equal to X, whereas in a normal programming language it would be necessary to provide an algorithm for square root before the expression was acceptable. A specification language is assertional, or non-procedural, as compared with a programming language or procedural. It is not clear whether or not there is any basic difference between these two types of language, or if, as is indicated plausibly by the example given above, an assertional language is merely a version of procedural language in which some procedures are left undetailed at the time of specification. This latter is equivalent to a constructionist view of mathematics. If this is the case, then the subsequent detailing of the undefined procedure, either by hand, or by computer, will provide a translation between the assertional and procedural form of the language.

### Language promulgation

Considering the importance of programming languages — and given the attention paid to other aspects of computing technology — far too little money is spent on their development and support.

The development and promulgation of programming languages is an extremely unsatisfactory affair. There is no commercial motivation to develop better programming languages. The hardware companies would not benefit, and software companies have neither the financial resources to carry out the development nor any method for taking a profit. This is a reflection of the generally unsatisfactory position of software as a saleable commodity, which has been brought about in part by the bundling policy of major manufacturers and in part by the inadequate provision for legal rights associated with programs. As a result, the development and selection of programming languages cannot be left to natural market forces — out there is no satisfactory alternative. The development of languages is usually carried out by academics or by standardisation committees, while promulgation is left to the efforts of institutional users or hardware manufacturers.

Academic study may suffice for advancing the art of programming languages, but it is hardly suitable for the detailed development of general language. Standardisation committees are even more inappropriate. A good language needs to combine simplicity, clarity and elegance — none of which is likely to result from the compromises inherent in the committee approach.

There are no adequate mechanisms for promulgating languages, although there have been some honourable *ad hoc* attempts, like the current CSERB support pattern for Coral. For such support to be effective, there is need for a central authority to disseminate information on a language and to provide training and advice; and there is no natural mechanism for funding such an activity.

In the short term, the situation with respect to language standardisation is unsatisfactory. To improve programming language to any great extent we need to develop and evaluate a wide range of language features, but it would be to our advantage generally to have a higher degree of standardisation than is presently available.

The concept of standardisation is ill defined. At least two levels may be distinguished:

- passive standardisation, where the form and detail of a language are codified, so that users can intercommunicate or interchange information, investment and skills
- active standardisation, where the use of certain languages is mandatory, or the use of other languages is forbidden.

There would seem to be little justification for active standardisation. Passive standardisation may, however, be too weak to ensure convergence on adequate language standards. It almost certainly needs to be supported by a positive programme of investment in facilities associated with the preferred languages, such as compilers and program-development systems, to ensure that the preferred languages do indeed present positive advantages to users during the early stages of standardisation.

Until there is a better understanding of language design there would seem to be little point in adopting new standard languages, since these can only offer marginal advantages, and these must be offset against the costs of increasing language diversity. A more attractive approach would be to make marginal enhancements to existing languages. Such enhancements should be obtained by removing limitations, or selecting more consistent subsets. Embellishment does not lead to language enhancement but merely to complexity. It might also be beneficial to de-standardise some of the less successful languages. Such steps would restrict the number of standard languages, and make the introduction of substantially better language easier in the future.

So far, standardisation has been discussed in terms of programming languages. Earlier, however, it was emphasised that programming is only one aspect of language usage, and that in the future other aspects of language will assume increasing importance.

Standardisation will be forced upon us by the need for computers to intercommunicate, rather than for the convenience of the programmer. For example, an increasing number of financial transactions will occur directly between computers: for this to happen, there must be agreement on the representation of numerical quantities and their significance. Similarly, as word processing develops there will need to be standardisation of addresses, orders and statements. Again, the Universal Product Code is an example of standardisation of meaning and representation, and there are a number of other such standards under consideration. In effect, all these are examples of language standardisation — and standardisation at a very strong level, since the *meaning* of the terms is standardised, and not merely their representation. There is a risk that *ad hoc* standardisation of this sort will pre-empt better languages, unless it is done within the framework of a coherent policy.

In the longer term a much higher degree of language

standardisation is inevitable, and it would be highly desirable to achieve a single standard language. This would facilitate inter-communication and allow the design of computer systems to to be optimised.

### The importance of language
*For the future, the U.K. is seen primarily in the user role. The U.K. has a good reputation in the language area. This would seem to be an area where large, but indirect, economic benefit could be gained by an intelligent programme of development and support.*

*Because of the failure of the market mechanism in the area of programming languages, there is a strong need for government involvement in both the development and the promulgation of language. It would seem reasonable for the funds for such an activity to be raised by a levy on users.*

*There is a need for a co-ordinated programme of research into languages, leading to a better understanding of the languages for the future. Much of this research is academic and would best be carried out in universities. There might be a case for estab-lishing a centre of expertise in languages to participate in and co-ordinate the programme, and to act as a centre for standardisa-tion in the future.*

*The attitude of the U.K. towards the continued support of Coral as a standard, and towards the EEC initiative on a new standard real-time language needs careful consideration against the need for non-proliferation of standard languages.*

*Present approaches to language standardisation are inadequate and need to be reconsidered. The U.K. might take an active lead in establishing a more appropriate method for establishing and maintaining future languages. Within the current context the U.K. should take a more active role to ensure standards in the best U.K. interests. This might include funding of participation and co-ordination of standardisation activities.*

## SOFTWARE TECHNIQUES
Software techniques are primarily concerned with the production

of software, rather than its use. As such their importance may decline as the degree of software replication increases.

## Operating systems

It is expected that the trend will be for computer manufacturers to fragment their operating systems and to bury them in the hardware. The reduced cost of processing and storage will enable more computing resource to be dedicated to the operating system, thereby reducing the pressure on operating-system designers to cut corners in terms of size and performance. The overall consequence of this should be that operating systems become better designed and more robust. In particular, the degree of instrumentation within operating systems should increase.

The consequences of these trends for the user could in some ways be unfortunate:

— There is likely to be less choice and variability in the facilities offered by manufacturers.

— By incorporating operating-system features in hardware, manufacturers will make it more difficult, both practically and legally, for plug-compatible companies to produce equivalent products.
In order to discourage the use of alternative products (as opposed to chinese-copy plug-compatible), manufacturers will needlessly complicate the interface between subsystems.

— The current boundary between operating systems and a language is arbitrary, and is not correct at a philosophical level. The boundary was determined by the difficulty that language designers experienced in formalising concepts like input-output and data structures. These boundaries will be made more rigid by the transfer of operating-system features to hardware, and therefore improvement will be inhibited.

— Because of the variability between manufacturers at the operating-system level, these developments will lock users into manufacturers more rigidly. They will also further lock in plug-compatible suppliers, who will concentrate to an even greater extent on IBM, giving added momentum to the 360 architecture as a *de facto* standard.

Overall, these developments are seen as leading to a greater rigidity of the computer market, making subsequent advances more difficult. If it had been possible to enforce international standardisation at the operating-system level for key characteristics like file interfaces, then the move to modularisation would be a big step forward. Such an approach, however, appears totally impractical at this time.

## Software developments

As explained on pp. 82-87, languages may be necessary, but are certainly not sufficient to improve the efficiency of software developments. The overall environment and management of software is of greater importance. Particular aspects are:
- convenience of access to computing facilities, particularly in terms of the user interface
- diagnostic and monitoring facilities in compilers, interpreters and run-time systems
- documentation facilities, particularly computer-based in terms of the proper documentation of programs and associated manuals, and in terms of the facilities to modify and update software in a controlled manner.

Many of these aspects can be promoted by the provision of the right computer facilities, but the primary essential is good management.

The long-term importance of software-development facilities depends on the way the industry develops. If there is a strong move towards standard software, as predicted in this study, then the importance of software-development facilities will decline, because they will not be the essential aspect of computing to the user. Under such a scenario, the important thing will be the improvement and standardisation of various aspects of the user interface, particularly the way the access is structured into information (files, etc.) and the way in which the user can modify information (text editing). It is considered that emphasis on these aspects could be more useful.

## Packages

The preceeding scenario implies that there would be much greater emphasis on the use of packages — that is, standard software sold in some machine-readable form, perhaps floppy discs or semiconductor cartridges. At present, virtually all packaged software is sold as stand-alone products. As the industry develops, there will be an increasing need for software packages to inter-communicate. This is the direct analogy of the requirement (already discussed on pp. 82-87) for independent systems to intercommunicate. A satisfactory solution will require the definition of high-level software interfaces and the definition of formal communication between these interfaces in terms of the semantic content of the information transmitted.

The requirement to implement package software at the micro-computer level (compilers, de-bugging aids, etc.) is already leading to changes in the design of microcomputers to ensure that the address space of the microcomputer can be used efficiently (since the package cannot be packed into a small address space) and to ensure that the program can be expressed in relocatable form. Similar developments may be necessary with conventional computer architectures.

### Conclusions
*The need for good software-development environments may decline as low-cost hardware discourages the production of bespoke software.*

*There is a need to concentrate on the improvement and standardisation of user interfaces, particularly in the areas of access and editing inormation.*

## COMPUTER ORGANISATION
There are two important questions to be asked about computer organisation:
- How will computers be organised to provide facilities comparable to present-day computers, but taking advantage of the improved technology?
- Will the present concept of the stored-program computer remain valid into the future, or will computing capabilities be deployed in different ways?

We proceed here to examine the first of these questions; the second is discussed on p. 99 ff.

## Low-performance versus high-performance technology

In the discussion on semiconductors on p. 47 ff., a sharp divergence was predicted between low-cost nMOS circuits based on optical processing, and high-performance circuits based on electron-beam processing, although no clear conclusion was drawn on the cost effectiveness of these two types of circuit.

The low-cost technology will provide processing capacity (processor plus immediate storage) equal to current minicomputers and to the low to medium end of the mainframe-computer range. Thus it must be considered adequate to cope with most of the common computer applications of today (e.g., payroll) at negligible cost. Under these conditions, the question whether high-performance circuits would offer economies of scale becomes irrelevant. Equally, it becomes irrelevant to consider sharing a processor between several tasks. It is more appropriate to dedicate the processor to a task and accept that at times it may be under-utilised.

This conclusion is significant, because it destroys a considerable part of the market for the super-processor, where it is used to reduce processing costs by economies of scale.

The more difficult question is whether the low-cost technology will displace the high-performance technology in the remaining super-computer applications. Here the requirement is a high throughput on a single program. The question is whether such problems can be partitioned into smaller units which can in some way be executed on a collection of processors, rather than being run as a single stream through one processor. Certainly, a significant class of large scale programs should be amenable to this approach, since they relate to models of three-dimensional physical systems, and can be formulated as array computations. This would further reduce the potential applications of high-performance circuitry within computing.

This still leaves open the possibility that high-performance circuits could achieve the required throughput more economically than a collection of low-cost circuits, or alternatively that there are some applications that demanding a throughput which could only be achieved by a collection of high-performance circuits. For the present, these questions are imponderable, because of the lack of information on the following:

- inneficiency caused by the partitioning of a problem among several processors
- performance potential of electron-beam technology circuits
- cost and availability of high-performance circuits.

Generally it would seem that low-performance circuits are adequate to meet the bulk of existing applications, and that high-performance circuits would only be justified if new large-scale computing requirements emerged. The discussion in Chapter 6 on the use of computers does not identify any such requirements.

## The federated computer

The most natural evolutionary development of the current mainframe computer would be to partition it into a number of co-operating computers, each performing an identifiable role. In a federated system of this type, one processor might be responsible for communications, another for managing the store hierarchy, another for the filing system, and yet others for running user programs, with a final processor controlling the overall scheduling of the system. In such a system, each processor is responsible for a specific operating function, and contains the relevant operating-system code. In all probability each processor represents the realisation of a virtual computer in the current operating-system context.

Where the operating system has been well structured, and there is not uncontrolled communication between operating system processors, or between operating system and user program, it should be relatively straightforward to evolve to this structure in a way fully compatible with existing user programs. In practice, operating systems are not this disciplined, so that the change may well involve substantial software replacement by the manufacturer.

Because of the degree of compatibility, it should be possible for manufacturers to upgrade users in stages, taking functions out of the existing computer and putting them into separate processors. It is expected that this could be the natural mode of migration for most mainframe manufacturers. The result for the manufacturer is that he has exploited the new technology and has an increased degree of flexibility in meeting the users' requirements when upgrading existing systems. The benefit to the user is that a high degree of compatibility can be maintained between old and new products.

The approach of providing a computer system as a federation of functionally dedicated processors will also be encouraged by the available technology, which will enable a microcomputer to be configured to a specific function, such as a system scheduler, by realising the control program as ROM storage on the chip. A further factor promoting this approach is the current unsatisfactory legal position over company rights to computer designs and software. By embedding parts of the operating system into microprogram, it would appear easier for companies to exercise control over the use or duplication of their software.

There are, however, two arguments against the federated-computer approach:

— It opens the mainframe computer to even greater competition from plug-compatible manufacturers, who may now have the opportunity to sell plug-compatible or alternative parts within a federated system.

— The range concept was valuable to the manufacturer, because it enabled him to exercise control over pricing, adjusting the price of various elements of a range to suit the market. This is no longer possible in the federated computer, since it is now practical to extend the capabilities of the system directly by adding the appropriate modules to match the increasing load.

For these reasons, although the manufacturers may adopt the federated-computer organisation as a method of implementing a computer range, the products may still be offered in a conventional format as monolithic computer systems, the modularity only being accessible to users in the long run as a result of competitive pressure from outside the mainframe companies.

The federated computer is also a natural development for the minicomputer manufacturer if he wishes to extend his capability up market. Indeed, the basic architecture of the minicomputer and the organisation of its operating system may well make this an easier task technically for the minicomputer manufacturer than it is for the mainframe company. On the other hand, the minicomputer company will be at a competitive disadvantage in terms of the software to exploit the system and in terms of marketing capability.

The federated computer offers the potential of faster throughput because the operations previously carried out concurrently (i.e., time-domain multiplex) in a mainframe computer can now be carried out simultaneously (i.e., space-domain multiplex).

The federated computer also offers the potential for cost reduction, even apart from any change in technology. This arises because it is made from a number of processors, identical except for the control program, thereby increasing manufacturer's volume for standard parts, and simplifying subsequence maintenance.

Finally, the federated computer has advantages in terms of reliability and resilience, since it may be able to operate to some extent even when one or more processors malfunction.

## The array computer

Another obvious way to exploit the availability of processing capability is the array processor. Here, the term is being used strictly in the sense that there is single instruction stream which is executed in parallel by multiple arithmetic units, usually organised in a two-dimensional array. Such systems are often called SIMD processors (Single Instruction path, Multiple Data path).

The attraction is that many of the current large-scale computing problems are directly amenable to solution by the array processor, since they are already formulated as sequential programs operating on multiple sets of data organised as an array.

As with current high-performance computers, it is likely that the array-processor design will be optimised to perform Fortran or APL calculations, rather than being designed as a general-purpose device. It is to be expected that each element of the array would be a processor with some local store — say, 1000 words — capable of performing floating-point arithmetic on internal operands, or on operands passed by adjacent array elements. In such a system, the overheads for providing a separate instruction path for each processor would be low, even though this facility is unlikely to be exploited in more than a trivial way.

The design of such an array processor using low-cost technology is straightforward in comparison with the problems posed by the present pipelined supercomputers. The comparative performance of such a processor would be extremely dependent on the details of the application. A basic-array operation time of one micro-second is feasible with current technology, and this will probably give a performance in excess of the 150 mega FLOP target for the Cray II on suitable problems.

The cost of the array processor should be low in comparison with current supercomputers, but the market is small. As a result, the array processor would seem to be an appropriate product for a small innovative company. It could well be that in the future such high-performance array processors will be sold as peripherals rather than being the preserve of the mainframe manufacturers.

### Arrays of microcomputers

In the federated computer and the array computer, collections of microcomputers are used in well-understood ways to implement systems where there is already an explicit degree of parallelism — systems which previously it has been impossible to use to advantage. There is also the possibility of using collections of microcomputers to solve problems which are currently formulated in a completely sequential manner. Two basic approaches have been proposed:

— That each microcomputer should perform a well-defined function in the system, say a procedure, and that the connectivity between the microcomputers should represent the connectivity between the functions of the system. This is analogous to plugging up the components in an analog computer to solve a problem.

— That the microcomputers should be connected in a regular array, and that the problem be mapped in some way on to this array. (Computer power by the yard: never mind the quality, feel the width!)

For any general-purpose system, the second approach is necessary, since the first approach would require some form of dynamic switching between the microcomputers, and this would have at least the same order of complexity as the components themselves. It is expected that problems will be mapped on to this regular array, thereby creating a virtual realisation of a system with arbitrary interconnectivity.

While there is no practical difficulty in building arrays of microcomputers (although current microprocessor architectures are not very helpful) — and indeed products based on this concept have already been offered — there are serious obstacles to their effective use:

- It is not known to what extent problems currently formulated in a sequential manner can be decomposed into a less sequential form.
- There are no adequate formal systems for representing these sequential computations, nor is there any theoretical basis for their study.
- The ideal communication properties between microcomputers are not known.
- The communication overheads between microcomputers are unknown: it is possible that for some classes of application these overheads would make the approach uneconomic.

There is, therefore, a great deal of theoretical and practical research to be done before arrays of microcomputers become an effective alternative. For this reason, it is expected that federated computers and array computers will become available earlier and will represent the dominant trend within the short and medium term.

A further factor operating against the array of microcomputers approach is the question of the extent to which there will be a market for such a product in the medium term. The single microcomputer should have sufficient competence to handle the bulk of projected computing tasks; the federated computer should be able to meet the needs of a central computing capability; and the array computer should satisfy the demand for high-performance computation.

In spite of the foregoing, the microcomputer array should be regarded as important, because it offers a rational approach to the provision of computer capability in the future, while the creation of an adequate theory for such systems could have fundamental importance for the understanding of computation and mathematics.

### Distributed systems

For the purposes of this study, a 'distributed system' is one in which a number of autonomous computers are interconnected by a communication system to provide capabilities for remote processing and remote access of information.

Distributing computing in this sense is already being used in an *ad hoc* way, and most manufacturers are developing network architectures — that is operating system commands and communication sub-systems enabling remote facilities to be accessed. Due to the lack of a general data-communications network, it is necessary for the network architectures to provide some form of communications system which can be implemented on the current voice-telecommunications network. As a result, a variety of network architectures are being developed with significantly different features.

In the long term, it is expected that most computing will have some requirement to access remote data or services, and that this will be done through a general data-communications network. To take advantage of this it will be necessary for computers from one manufacturer to be able to access data or services from computers provided by other manufacturers — the so-called 'open-network' philosophy. Thus the present situation, with a variety of network architectures, is seen only as a temporary arrangement. In the long term, the manufacturers will find it necessary to adopt the protocols of the standard data-telecommunications network, and these protocols will impose a much greater degree of standardisation on the internal characteristics of a computer system in order to achieve the necessary capability for interchange. Areas where uniformity will be required are job-control languages, file naming and format, and process structuring. The route to such standardisation will be through the definition of data-communication protocols such as that already in progress through the X25 family from CCITT. It is important that these standards be framed with adequate understanding of the limitations they will place on computing in the future.

**Conclusions**

*The demand for large-scale computing, based on economy of scale, is expected to decline.*

*The federated-computer approach enables many aspects of the technology to be exploited by manufacturers in a straightforward way, while allowing a high degree of compatibility with existing architectures. This is to be expected to be the preferred direction of development in the medium term.*

*The array computer has a well-defined but small market, and may well exist as a peripheral rather than as a free-standing product.*

*Communication protocols are likely to have an important effect on computer organisation, and need to be framed with due attention. A major effort is required to ensure that the X25 family of protocols is adequate, and to investigate the further development of communication protocols.*


## TECHNOLOGY DELIVERY

The past twenty-five years have established a pattern where processing capability is delivered in the form of the general-purpose programmable computer, which is then tailored to specific applications by the provision of custom software, either by the user or by a system house. To assist in the tailoring, there has been increasing provision of such standard software as programming languages, operating systems and utilities, and even of standard-application software which can be adapted to specific users.

It does not follow that this will be the normal method of delivery in the future. The changing cost of the technology, the changing uses and the growth of use of the technology might make other methods more attractive.

### Microcomputers

The semiconductor processing capability described on p. 47 ff. enables a pattern of active circuits to be created on a semi-conductor surface. Potentially, such a capability could be organised in many ways but, as explained on p. 75, it is expected that the use of program techniques will rapidly predominate over

99

the use of special circuitry. The primary reasons for this are:
— That the programmable microcomputer represents a more efficient use of circuit area than does a random logic.
— That the design techniques for using a microcomputer will (eventually) be far simpler and more efficient than the design techniques for circuitry.

## Programming

It is not realistic to expect the cost of developing software to fall significantly. The use of computer aids like programming languages and better development environments can reduce the mechanical aspects of programming, and the falling cost of computer equipment means that the capital investment required to develop programs will fall. There is, thus, considerable scope for a reduction in the cost of developing software (say, by a factor of five); however, this is not significant when compared with the fall in the cost of using hardware. Further reductions in the cost of developing software would seem unlikely, because the effect of the various computer aids is to make the design of software not easier but harder. At present, much of the programmer's time is spent on mechanical aspects of programming, such as mapping the problem on to a specific computer configuration. It is reasonable to expect that these mechanical aspects will be automated, and this will leave the programmer with the hard part, which is to express and solve a problem in formal terms — something directly akin to pure thinking.

Thus, so far from the skills required for programming being reduced, they are likely to increase as the routine parts are automated, leaving the programmer as a pure problem-solver in a particular domain. What technological developments will do is reduce the need for the programmer to have any specific expertise in computing, so that creative individuals in various intellectual domains will be able to use the computer directly to solve problems for themselves and others.

The elimination of routine programming has obvious implications for future personnel requirements in information technology, and needs to be considered carefully in planning future educational policies for this area.

**Programmed products versus programmable products**

In many areas, the microcomputer will be delivered as a fully programmed product — for example when it is used as the control element in a consumer product. It is not so obvious how the market will develop in the areas where the microcomputer is used as a pure information product. There are two basic alternatives:

— that information products will be delivered as programmed devices for specific applications

— that information products will be delivered in two parts: as a general-purpose device (which for convenience will be called a *puter*, since it may be rather different from a conventional computer), and as programs to make the computer perform a specific function.

The arguments in favour of the first route are that by packaging the software as a complete product the manufacturer provides greater protection for a rather vulnerable asset, and that this approach adds more value so is commercially more attractive. The argument in favour of the second approach is that most information applications need nearly identical hardware (an alpha-numeric display interface to the user, and a keyboard input), and duplication of this facility for each information product would be unnecessry and costly.

From the manufacturer's point of view the first approach is far more desirable, and until there is a standard puter there is little alternative. Thus in the short and medium term there is expected to be a rapid growth in the number of programmed information products becoming available. In the longer term, the advantages of a standard puter would appear to outweigh the commercial pressures, and it is anticipated that a relatively standard human-interface puter will be developed, which will be able to perform different functions with different plug-in software modules. Given the declining costs of microelectronics, such modules would almost certainly be semiconductor, and might well contain their own processor rather than sharing the processor(s) in the puter itself. The form of the puter is arguable; perhaps the electronic typewriter described on p. 145 would be a suitable basis.

If software is to be delivered in the form of silicon, it may well be that the semiconductor manufacturers will emerge as the software publishers of the future, and that the primary function of the software to them will be to create a market for their semi-conductor-processing capability.

## Software

There are widely varying views about the future importance of software, the two extremes being summed up by these two catch phrases:

- 'If it is so expensive to develop the software, let's do it in hardware.'
- 'The hardware will be so cheap, it will be given away with the software.'

Paradoxically, there is some truth in the first view:

- Engineers are accustomed to much stricter design constraints than programmers. The application of techniques like drawing-office documentation, design-modification procedures and enforced modularisation (e.g., to fit a standard printed circuit board) can greatly improve the development of software.
- The removal of constraints on efficiency and size in software can lead to better disciplined designs and better structured software, since there is less need to break the rules.
- The lower cost of hardware makes it more economic to tackle simple applications, which have less demanding software requirements.

For all these reasons, the software problems of using micro-computers may be reduced in comparison with the software problems of conventional computers.

The second view, often expressed by software houses, has an obvious derivation. The cost of hardware has already fallen to a level where the cost of writing software dominates in many conventional computer applications. Already, the store to *hold* a program costs 10,000 times less than the cost of *writing* the program and, even given the improvements in programming efficiency suggested above, the imbalance is likely to get worse.

The difficulty with this argument is that there is a very obvious way to reduce the cost of the software — by selling multiple copies. As already indicated, the physical cost of delivery is negligible in comparison with the front-end cost, so that amortising these costs over multiple copies can lead to a dramatic fall in the cost of software. The impact of this can already be seen with a programming language like Basic. System software of this sort is already sold as a product by the computer manufacturers, and typically an acceptable price for Basic on a minicomputer has been around £1,500. Recently, Basic has become available as a product for the hobby market, and the typical selling price is now around $15 for a comparable product.

The only obstacle to this approach is the user. Will he be prepared to accept standard software, or will he continue to demand bespoke software to meet his individual requirements? While the cost of hardware remained high, the use of bespoke software provided useful advantages, but it is not clear what differential the user will be prepared to pay to maintain these advantages. Certainly, while the differential is a few per cent, the balance is not obvious; but the use of microelectronic products with standard software has the potential to cut computing costs in business data-processing by factors of ten or more. Under these conditions, it would seem unlikely that the user would choose to go down the bespoke route.

For the present, the software industry tends to see the advantages of the standard software product, but not its disadvantages. The advantage is that the development costs can be spread over many copies, so that under the present pricing regime large profits will be possible, and indeed are likely in the short term until the market recognises the position. The disadvantage is that, thereafter, the nature of the software industry changes and becomes dominated by the need for volume.

In this second phase, expected to develop in the medium term, the cost of developing software is extremely high, since it needs expensive capital equipment; but the delivery cost is low. This is exactly the pattern of the semiconductor industry, and it might be expected that the software industry would develop the same competitive charactistics. These are:
— the strong relationship between volume and price, leading to aggressive pricing policies as manufacturers strive to achieve volume: this places emphasis on the financial

103

strength of manufacturers and can lead to very low profitability

— the establishment of *de facto* standards — products which because they have a volume market achieve a low price and prevent the entry of competitors, even though the competitors may have technical advantages

— the learning-curve phenomenon, where the price of a product falls rapidly as volume is established, leading to great pressure to be first in the market with a new product, and hence to a high rate of innovation.

There is a third phase which is to be expected in the long term, and could emerge rather sooner. With the declining cost of computing, the capital requirements to develop computer programs will fall, and the most important characteristics become the skill of the individual programmer and the ability to market effectively. These are the characteristics of the book-publishing business, and it is reasonable to expect that the software industry will develop along similar lines — with a few companies responsible for publishing and marketing products which are created by individuals with the necessary talent, who, like authors, could well operate on a freelance basis. This pattern has already emerged in the hobby computer market, where program-publishing houses exist; and Byte magazine also publishes programs.

### The price of software

It is instructive to consider the pocket calculator as an example of an information product. The difference between a simple four-function calculator and a multi-function calculator lies almost totally in the complexity of the software. When these products were first introduced, there was a very large differential between the prices that the two products could demand. However, under competitive pressure, the price of the multi-function calculator has fallen to a level where it can only command a small margin over the simpler product.

Another interesting aspect of the calculator market has been the emergence of the clock calculator. To provide a wide range of clock functions, a reasonable interface is necessary with the user, and this implies a keyboard. A calculator on the other hand only requires the addition of a precision crystal and it can function as a clock. As in the case of the multi-function calculator, the marginal cost of providing the additional facility is small, so that two products are likely to converge. In the future it must be expected that most calculators will provide a much wider range of user functions and have the ability to tell the time. It appears that, in the future, powerful software products will drive out trivial software products, because the marginal cost to achieve the powerful product is small.

Once again the analogy with books would appear to hold. The retail price of a book is calculated from its unit cost, which consists of creative and editorial costs, paper, printing and binding, the delivery costs and the retailer's margin, all divided by the number of copies produced. The content and quality of the book need have little influence on the price. If the more highbrow books tend to be priced more highly, that is because few people buy them — not because such books are, *per se*, more expensive to produce or more highly valued. A 'good' book may be more profitable than a bad book, but that will only be because of the number of copies sold, not because greater monetary value attaches to higher quality of content.

A closely analogous situation may be expected to pertain with software in the future. The price of software will be closely related to the price of the delivery vehicle, which will almost certainly be silicon. It is also likely that the average user will continue to find difficulty in discriminating good software from bad — so perhaps there will be a role for program critics in the electronic newspapers of the future.

## Conclusions

*The microcomputer is seen as the primary information product at the semiconductor level.*

*It is anticipated that there will be a continuing division between hardware and software for information products, but that the nature of both will change.*

*The standard hardware product will not be the computer, but the interface device to the human user. This could well take the form of the electronic typewriter described in Chapter 6.*

*The changing cost of hardware is likely to have a radical effect*

105

on the nature of the software industry, which could move towards an author/publisher structure.

The price of software is likely to be independent of its value and dependent instead on the delivery mechanism, which is most likely to be silicon. The semiconductor companies could emerge as the software publishers of the future.

The elimination of the routine aspects of programming will change the nature of the task, and this needs to be considered carefully in planning future educational policies for this area.

# CHAPTER FIVE

# THE STRUCTURE OF THE COMPUTING INDUSTRY

## CURRENT ORGANISATION OF THE INDUSTRY

All too often, the computing industry is equated with the mainframe manufacturers. In part this is due to historical circumstances, but this view is perpetuated by the visible nature of mainframe computers and the relative ease of collecting statistics about them.

### Structure of expenditure

Regrettably, statistics on the computing industry are not good, and any figures must be treated with considerable caution. An approximate breakdown of every pound spent on computing in 1975 was the following:

| | |
|---|---|
| data-processing salaries | 30 |
| mainframe computers | 35 |
| minicomputers | 10 |
| software and services | 15 |
| data communication | 5 |
| supplies | 5 |

As in any such breakdown, apart from the uncertainty of the base data, the categories themselves are imprecise. In this case, mainframe computers are taken to be systems costing more than £50,000 and minicomputers to be systems costing less than £50,000, so that the latter category includes office computers and products like System One from IBM. Data communication is taken to include charges to the BPO, and associated modem equipment.

Expenditure on data-processing salaries is a declining proportion, and this decline is expected to continue. The decline is partly a consequence of the growth of the minicomputer — which has a lower proportion of directly attributable staff — and partly a consequence of the growth of various interactive uses — where the immediate uses of computing are no longer accounted for under a special budget heading within company operations.

The mainframe has been representing a steadily declining proportion of expenditure, and the statements by IBM to Justice Department that its share of the total computer market is falling would appear justified. Because of the expense of the equipment, the mainframe market is characterised by well-defined customer population. This is largely penetrated, and further growth must come from expanding the use of computing within the established range of users. This means that the concept of customer base has become very important, and the main competitive tactic left to manufacturers is to attack the customer base of their competitors.

There has been considerable concentration in the supply of mainframe computers. One company, IBM, dominates the market with over 50% of the revenue. The number of general competitors in this market has declined steadily over the years, until there are fewer than ten other companies of any significance. This decline can be directly attributed to the impact of technical change and the need for massive investment to maintain a fully competitive range of products. There has, however, been a growth in the supply of plug-compatible equipment from companies who supply sub-systems compatible with the products from a mainframe manufacturer (usually IBM). In contrast to the companies offering non-compatible mainframes, the plug-compatible companies have enjoyed considerable success. They have been growing rapidly and threaten to change the nature of the mainframe business. Market projections show the overall mainframe-computer sector

108

growing slowly over the next few years at 8-10% p.a., with the plug-compatibles growing more rapidly within that sector.

The majority of the plug-compatible companies have evolved from peripheral manufacturers. Traditionally, the peripheral industry was captive to the mainframe suppliers and was largely printer-, reader-, tape- and disc-mechanism-oriented. As scale grew, independent suppliers entered the market, and most non-US mainframe suppliers increasingly depend on consortia of US-dominated peripheral manufacturers. The development of computer ranges plus the increasing degree of interface standardisation, itself promoted by the trend to communications, resulted in the explosive phenomenon of the plug-compatible industry, and today the peripherals industry should be regarded as part of that plug-compatible industry. The development of this industry is playing a critical part in the price erosion of the mainframe-computer business under pressure from companies like Itel, Amdahl and Telex. A noticeable feature of the latest plug-compatible offerings is the trend towards microcomputer-controlled peripherals and the introduction of plug-compatible processors which exploit the latest semiconductor technology.

The minicomputer industry was started by selling basic hardware in situations where little support of software was necessary. As the volume has developed, it has been possible to provide increased software capability while maintaining low prices, so that the market has expanded to encompass various forms of business computing. As a result, the original minicomputer companies have been joined by office-computer companies and by the mainframe manufacturers in competing for low-price systems. The growth of the minicomputer market is still sustained around 30% p.a., primarily by activity in the business sector.

Forecasts of the future growth of the minicomputer business are very variable. This is in part due to the differing definitions of this sector; but it also represents uncertainty, first about how far the minicomputer can displace the mainframe in distributed systems, and secondly about the extent to which the minicomputer business will be eroded by the microcomputer. Forecasts can vary between 5 and 40% per annum in the short to medium term. The consensus view would probably put the growth of the sector considerably above the growth in the mainframe sector at, say, 20-25% p.a.

The state of affairs on the software and services sectors can also be misleading. This is because it consists of three businesses, with quite different financial characteristics:

- bureau services for batch and time sharing, which are capital intensive
- consultancy and software, which are labour intensive
- turnkey sales where the bought-in content inflates turnover.

Overall, there has been a rapid growth in the services industry. This has been stimulated by the emergence of low-cost computers, where the user requires external assistance to augment the hardware capability he has bought and to avoid the need to acquire in-house expertise. World wide, the largest companies are primarily bureau services, with American companies like ADP, TRW and McDonnell Douglas having turnovers well in excess of £100m p.a. The scale of U.K. companies is small by comparison with those in America, or in Europe, only seven companies having a turnover in excess of £5m p.a. Forecast growth for the service industries is generally strong at 30% p.a., with some dissension. The variability is caused by the question whether or not the bureau operation will be impacted by the availability of low-cost business systems. In practice, the revenues of a large bureau like ADP are primarily derived from batch processing for very small companies who each contribute less than £1500 per year, so it is unlikely that this business will be impacted in the short term.

The cost of data communications is already high and is growing rapidly with the increased demand to intercommunicate information between computers. On present pricing trends this could rapidly become a significant proportion of total expenditure on computing.

### The U.K. situation

The level of investment by the U.K. in computing appears quite good in relation to other countries. On the basis of various 1975 figures we have:

See Table on opposite page.

110

TABLE: Various Measures of National Investment in Computing 1975

| | Investment in computing, in £M p.a. | Population in M | GNP in £B p.a. | Investment per capita in £ | % GNP invested in computing |
|---|---|---|---|---|---|
| U.K. | 910 | 55 | 85 | 16 | 1.1 |
| Benelux | 300 | 25 | 60 | 12 | 0.3 |
| France | 910 | 50 | 145 | 18 | 0.6 |
| Germany | 980 | 60 | 185 | 16 | 0.5 |
| Italy | 380 | 55 | 70 | 7 | 0.5 |
| Japan | 1200 | 110 | 215 | 11 | 0.6 |
| US | 7800 | 210 | 645 | 37 | 1.2 |

Source: OECD Publications

Per capita, the U.K. level of investment is comparable with that of other industrialised countries, although far below that of the US. In terms of the proportion of GNP, the level is above that of the other industrialised countries and matches the US.

On global terms, the U.K. computer industry is reasonably in balance with the total market. The U.K. represents about 5% of the world market, and in both the mainframe and software-services industries the U.K. has about a 5% share. Its weaknesses are in the areas of minicomputers and plug-compatibles. It is arguable that these weaknesses can be traced back to the lack of support at a critical period in the development industry and to the strong protectionist attitude towards ICL which has prevented the development of the plug-compatible industry for ICL products and inhibited the development of plug-compatible products for IBM equipment.

The DoI trade figures show a substantial flow of goods in computing, as is typical of most sectors of the U.K. economy. Officially, there has been a trade deficit of about £50m p.a.; but this may be more a figment of the figures than a reality, caused by double-counting peripherals. In all probability, the real figures are in reasonable balance. Because of the high foreign content in its peripherals, ICL is a substantial importer, and may make less of a contribution to the balance of trade than some American companies like IBM and Honeywell (this, of course, ignores the export of profits).

The software-and-services industry has a turnover of about £250M p.a. — that is, about one third of the hardware market. The level of imports and exports is negligible (less than £10m p.a.), and is reasonably in balance.

**Conclusions**

*The organisation of the computing industry is more complex than is often assumed in government policies. This organisation is not fixed, but is changing rapidly. Within the industry, the mainframe computer is declining in importance.*

*The position of the U.K. as an investor in computing is better than is often assumed. However, because of our low level of*

112

*productivity, we can only just match the level of investment of
other industrialised countries.*

*The U.K. computer industry, both hardware and software, is
in good balance with the U.K. use of computers. The two areas
where it is weakest are in minicomputers and plug-compatible
equipment — both areas where major growth is expected.*

*Information and statistics on the U.K. computing industry
are poor. There is a good case for a study to establish a datum
point for future planning. Such a study should particularly
examine the balance of trade of the industry as a whole, and the
contribution of individual companies.*


## MECHANISMS CONTROLLING STRUCTURE

Various parts of the computer industry show a strong tendency
to monopoly. This is primarily caused by the need for compatibility
across a range of products.

### Product monopoly

The most noticeable feature of the computer industry has been
the tendency to monopoly. The dominant position of IBM has been
well publicised and discussed. The dominant position of Digital
Equipment Corp. in the minicomputer market is less often
remarked. It is clear, however, that Digital occupies a very similar
position — in market share, in the price premium it can command,
and in its competitive and restrictive strategies.

Various reasons have been put forward for this tendency to
monopoly, primarily based on arguments of economies of scale
and excellence of management. There are some economies of
scale in manufacturing, and perhaps even larger economies in
selling. There is little evidence of exceptional management;
consider, for example, the IBM errors over unbundling, or 360
technology, and the failure of Digital to identify the 16-bit market.

The primary reason for dominance appears to be the phenomenon of product monopoly, brought about by the modularity
characteristics of the computer. A computer is not a single
product, but a range of products which must be completely compatible. Each customer buys a system, which, to a first approxi-
mation, is a random selection from this range. Consider what

happens if a customer who wants a particular configuration goes to a manufacturer whose range is incomplete. Each item he wants has only a probability of being available, so his probability of being able to buy a system with n components is only pn, a number which reduces as n becomes large and p becomes smaller. For example, if p is 0.8 and n is 4, the chances of a suitable configuration being available are only 0.4. The situation is not quite so bad as this simple model suggests, since particular types of customers will make correlated choices, while many elements of the product range will be minor variants so that to some extent they are interchangeable (for example, a matrix printer for a daisywheel printer).

The effect of this phenomenon is that if one manufacturer is able to offer a larger product range than another, his competitive advantage is multiplied disproportionately. The size of the product range is related to turnover, since the limiting factor is the high investment needed to create new products. Thus, if one manufacturer establishes a size advantage over another, he can readily maintain or extend this advantage by enlarging his range of compatible computer products.

So far, five competitive strategies have been employed to overcome this market characteristic. Three have been unsuccessful, and two have met with success:

— Direct competition by matching product and market capability. This is the basic route taken by the mainframe manufacturers, of whom ICL is a good example. This fails because it does not take into account the basic mechanism leading to market dominance.

— Competition concentrated in a limited market sector. For example, Burroughs in banking, Univac in real time, CDC in large computers. This approach is better, because the manufacturer can exploit the correlation between customer choice to provide a more limited product range which has a higher probability of satisfying customer requirements in the selected market. It has not been very successful, because a considerable part of the product range is common to the majority of applications, and restricting the market area means that the revenues generated to support the common part of the product range decline rapidly, so that the company cannot support a fully competitive product range even within a limited market sector.

- Direct competition by providing a compatible product range. This was the route taken by RCA, with disastrous consequences. It is rather worse than the first option, because the deficiencies in the supplier's range are more directly apparent to the customer.
- Identification of a market in which the dominant company cannot compete. This is the route taken by the mini-computer companies. The mainframe manufacturers cannot compete with their standard products because they are too expensive, and have been inhibited from developing competitive ranges because of the risk that these will destroy their established high-revenue markets.
- The sale of components within the dominant product range. This is the approach of the plug-compatible manufacturers. This competitive technique has been highly successful, firstly for peripherals, then for software and more recently for electronic storage and processors. It avoids the primary difficulty of general competition and the need for an all-embracing product line, but can only be successful if there is an established dominant product line, and sufficient modularity for the customer to be able to purchase from multiple sources. The success of the plug-compatible technique has been largely unrecognised in the U.K., because its operation has so far mainly been restricted to the US. Its importance is attested by the strenuous efforts that IBM, and more recently Digital, have made to discourage plug-compatible competition. Paradoxically, while plug-compatible competition may threaten the revenues of the dominant manufacturer, it makes his products more attractive to the customer, who now has a degree of choice. As a result, the dominance of the product line is reinforced, leading to the creation of a product monopoly in place of a manufacturer monopoly. In the case of the IBM system 370 range, it is now possible to purchase a complete 370 system which does not contain any IBM product. As a result, it is

extremely improbable that IBM could introduce a new product line that was not fully compatible with the 370, without causing mass migration of its customer-based plug-compatible manufacturers offering enhanced 370 products.

The mechanism of product monopoly explains the current characteristics of the computer market and enables positive predictions to be made about the way in which the structure of the market will develop.

## The U.K. industry

Although the U.K. computer industry has been able to survive and maintain a reasonable share of the world market, it has not been outstandingly successful. The success of U.K. nationals at both the technical and the marketing level in US companies suggests that this is not due to a lack of basic talent. There would seem to be at least three structural reasons which make it difficult for U.K. companies to be competitive:

— The small size of the U.K. market, which is inadequate to support the scale of investment necessary to create competitive product lines. U.K. companies must, therefore, export — which adds to their costs and to the difficulty of launching technologically complex products.

— The U.K. market is slow to accept new technology. As a consequence, computers can become established in the more advanced US markets, and when the U.K. market opens, can enter with a product which is proven and whose front-end costs are partly amortised.

— The growth of the plug-compatible market, while it may impact the sales of the primary manufacturer, greatly reinforces the competitiveness of his product line, since it gives the customer a choice of supplier. Under such conditions it becomes more difficult for non-compatible suppliers to compete. Most U.K. companies are not plug-compatible, but are competing in markets where plug-compatibility is becoming increasingly important.

Any support or promotion of the U.K. industry must provide a specific solution to each of these structural difficulties.

116

# CHAPTER SIX

# THE USES OF INFORMATION TECHNOLOGY

This chapter reviews various uses of information technology. In general, these cut across industry sectors, so each class of use will be relevant to many industries.

Each class of use imposes various different requirements on the underlying equipment (hardware and software) and it may be expected that there will be greater differentiation of equipment against use in the future. This is most clearly seen in the basic units of information that the equipment has to handle. Thus, for research, the basic unit is a floating-point number, whereas for data processing it is an accounting unit for which a 32-bit fixed-point representation is adequate. Similarly, the unit required for office automation is the alpha-numeric character, normally represented by using the 8-bit ISO format, while the unit for telecommunications is the bit itself. These classes of use are differentiated in other ways — system characteristics, reliability and cost — and this could lead to a greater divergence in the types of information equipment in use in the future.

The various classes of use have been arranged approximately in order of the date at which they became significant. Thus, the initial use of computers was for research and for military applications. At that time, the potential for the use of the technology was not recognised, and a number of forecasts were made that all the computing requirements in the U.S. (or the U.K.) could be satisfied by a handful of computers. These forecasts were overturned primarily by the recognition that computers could be applied to commercial data processing, to the extent that commercial data processing is now the largest single use of computing. It does not follow that this will remain the dominant use in the future. As the cost of the technology falls, new types of application, like office automation, become possible, and these may well come to dominate the market in terms of value. There is a great danger in looking at information technology in terms of the current pattern of use. The dominant pattern of use has changed in the past and may be expected to change again.

## RESEARCH AND DEVELOPMENT [R & D]

R & D was one of the earliest applications for computing. Nowadays most R & D is dependent, at least to some extent, on computing and many important recent developments have only been made possible by computing. Because of the current high level of usage, future developments in information technology are not seen as likely to cause any dramatic change in the pattern of research, although the role of the computer will become even more important.

### Use in research

The R & D phenomenon in general is only recent. Big corporate research was built up in the 1930s, and great emphasis was laid on central-government research after the two World Wars. It is generally accepted that the increasing complexity of advanced technology, and the scale of equipment, will continue to encourage centralisation. In many cases it is the specialised computing facilities (hardware and software) that are becoming the focus for research teams.

The growth of interactive computing may be expected to improve the process of creative research. With an interactive capability, a scientist can evaluate a theoretical formulation, gaining understanding of its properties and behaviour. Such an

118

approach, illustrated some years ago by Codd's work on cellular automata, enables alternative theories to be searched rapidly and critical experiments to be identified.

The use of large-scale computing (particularly array processing) for modelling phenomena with inherently local interactions which propagate through a two-(or more)-demensional space — for example in meteorology or in nuclear physics — offers the opportunity for development in areas not directly amenable to mathematicl manipulation.

Finally, the computer might be expected to provide more rapid dissemination of results. Propagation of scientific knowledge is, in many respects, very inefficient, and the ready access of many scientists to computer networks will promote the trend, which is already apparent, to replace journal publications by circulation of results in electronic form. This is likely to be one of the earliest types of the electronic services which are described on pp. 154 ff.

## Use in development

Development covers a wide range of practical activities, concerned as it must be with a wide range of process technologies for alternative-product realisation. It is being realised increasingly that functional analysis of a process or product is relatively invariant to technological implementation — a concept somewhat imperfectly enunciated by Zwicky in *Morphological Analysis*. One use of computing power will be to extend such functional analysis using computer-search capabilities to identify new end-products and alternative technological realisation.

Design, as well as using conventional computing approaches, will increasingly use interactive terminals to search for standards and data, to capture design concepts, to explore alternatives and to specify the design itself. The concept of interactive design, via a simulation approach, will undergo a change, with computers increasingly being used to harden (through material or component choice) the more free-ranging expression of the designer. Automatic drawings, increasingly to be regarded as mere human interpretable checks on data whose fundamental representation is electronic, will tend to move to semipermanent display media, using visual terminals until direct electronic recall is fully established. The advantages of complete electronic representation of

119

design include the ability to update and amend designs, to generate consistent documentation and to provide outputs in a form compatible with manufacturing systems based on computer procurement and planning.

## Conclusions

*The development of computing capability will increase the efficiency of the various activities covered by the R & D spectrum. This could move the balance back towards the creative scientist and increased job interest.*

*The R & D market represents a rapidly declining proportion of the total computing market, and R & D users must increasingly expect to use computers designed for other markets, making any necessary adaptions to the software and systems.*

*Adequate investment in computing facilities will remain an essential prerequisite for effective R & D. This is a prime example where the use of computers can generate great net benefit and should not be prejudiced by attempts to support the computer industry itself.*

## MILITARY APPLICATIONS

The first computers were built for weapon design and cryptography. Since then, the computer has been an important aspect of military hardware, and a major aspect in the balance of power between East and West.

In the past, military investment has played a large part in the development of computing. Microelectronics as a whole is now an essential element in strategic capability. With the development of the low-cost microcomputer, military and civil needs are diverging, and indeed can come into direct conflict. Many of the developments in the technology are upsetting the current balance of power and thereby only adding to the current instability.

### Impact on applications

Conventional warfare is currently undergoing reassessment following the high attrition rate caused by precision-guided missiles in the last Arab-Israeli conflict. As a result of microelectronics the balance of advantage in such conflicts has now swung to defence, since attack and ground conquest require mobility and therefore exposure. Further development in micro-electronics will increase this imbalance.

Nuclear war is potentially being destabilised by the current development of the cruise missile. In this case, the trend to microelectronics and the relative ease of copying may well so complicate the strategic assessment through rapid proliferation that the inherent stability of the two person negative sum game between the major powers may vanish.

The relative untraceability of a nuclear-armed nuclear sub-marine is to some extent being eroded by new detection methods using satellites and fixed detection arrays on the seabed.

In the field of communications, the improved capability of com-munication systems and the ability of computers to provide secure encypherment has reduced the vulnerability of geographically distributed forces.

Warfare using the space and time extension afforded by stationary-surface satellites seems too powerful to be ruled out by treaty. In unconventional warfare, developments have been suggested where microcircuit control and capability may well act as accelerators. Various near-space arrays are now reaching feasibility with actions as diverse as clearing windows through protective atmosphere for extreme outer-space radiation pene-tration, or complete blanking of local EMS-based communication.

Terrorist activity is encouraging the development of surveil-lance-, data-collection- and data-correlation- systems. Unless agreed upon rapidly, such networks may be developed and misused in a police-state sense, thereby achieving the polarising objectives of many ideologically based terrorist groups. The availability of microelectronics makes terrorist weapons both more dangerous and easier to develop. In this sub-war area, still on the increase, microelectronic ability is likely for some time, therefore, to have the effect of increasing the stakes in the terrorist/counter-terrorist game.

### Balance of power
The provision of an advanced microelectronics capability by the US has been an important factor in the balance of power between East and West. To exploit this advantage, restrictions have been

made on the export of advanced electronic systems to the Eastern Bloc, as for example in the recent case where export of a Cyber 76 for weather forecasting in Moscow was refused.

The development of low-cost microelectronics, and in particular the microprocessor, is making this policy largely untenable. It is clearly impossible for the US to control the distribution of components costing only a few dollars, and there can be little doubt that the Russian answer to the cruise missile will use American microprocessors as the control element. This, of course, was also possible in the past; however at that time, the lower complexity of the circuits themselves created a barrier, because substantial quantities were required to create the control device, while additional knowhow and skills were also required, so that the diffusion of components was less important.

One possibility for the future would be for the US to restrict access to advanced technology, not allowing this to reach the commercial market place. This policy could be attractive to the US because many of the more interesting military applications require performance in excess of that achievable by current technology, while the majority of envisaged commercial applications could be satisfied using the current technology.

### Conclusions
*Microelectronics capability is the key to modern military power. To be effective militarily, a nation must have access, direct or indirect, to the latest semiconductor technology.*

*The development of the microprocessor has largely destroyed the barrier erected by the US to prevent diffusion of control technology to the East, and this must change the balance of capability between the two major powers.*

*The use of information technology is changing many of the assumptions on which the current balance of power is based, and this must be a destabilising influence.*

### DATA PROCESSING
The pattern of development in data processing has been greatly constrained by the cost and capability of computing equipment. Although the determining factors are now changing, the high

level of investment in existing equipment and practice, together with the vested interests of suppliers and users, will make this market slow to respond to the opportunities of the technology.

## The development of data processing

The cost of computing has been high and its technical complexity has been great. This has had four effects:

— The benefits of computing have been restricted to the larger companies who could make the necessary capital investment. This has given these larger companies a competitive advantage over smaller companies and has been one factor in the growth of the large corporation.
— The computer has promoted the trend towards centralisation. This has often meant that the introduction of data-processing techniques has required radical alteration to company practice.
— The application of data processing has been restricted to certain classes of transaction within a company, where the high cost per operation could be justified.
— Data processing has emerged as a separate profession, often wielding considerable power within a company organisation.

The development of low-cost computing has the potential to reverse each of these effects, with profound implications for data processing as it is practised today.

The reduced cost of hardware offers the possibility of reducing the cost of data processing. Such benefits will only be obtained if:

— The cost of software is reduced. This means a move from custom software to standard products, perhaps with the possibility that the customer can adapt the product to his requirements.
— The cost of introduction is reduced. This can be achieved by the use of standard products (software and hardware) which reduce the selling cost, and by the use of products which are more closely related to current business practice, thereby reducing the need for retaining and reorganisation.
— The cost of operating the data-processing system is

123

reduced. This means the elimination of specialist staff, including programmers and operators.

All of these actions are possible and the industry is gradually moving towards a product orientation. At present it is in an intermediate stage, with the low hardware costs often masked by high selling costs and partially customised software. As a result, although the minimum investment cost for an effective data-processing facility has fallen, it is still in the range £20K-£50K, restricting data processing to companies with a turnover of the order of £1m a year or more.

The basic configuration for data processing can be reduced to a visual display unit with programmable capability, a storage medium (which at this stage might be floppy disc, but which in the future can be expected to be some form of semiconductor cassette) and a minimal printer. In hardware terms, such a system could be delivered for £2.5K so that the acceptance of software products would greatly extend the potential market for data processing.

The use of a more interactive mode of working can also reduce the cost. By exploiting the intelligence of the operator, it is possible to reduce the degree to which the system must be programmed with all the alternatives, and thus enable the operator to match the facilities of the computer to existing manual systems. Finally the interactive system can be exploited to train the operator directly — and to train future replacements.

The technology also changes the arguments for centralisation. There are no longer the same cost advantages in centralising processing, although cost advantages may remain for centralised storage. This means that, in the larger company at least, it is possible to consider the provision of computing capability where it is required within the company, rather than reorganising the company to centralise its computing. The essential merits and demerits of distributed computing are obscured by the vociferous support of an interested group — namely minicomputer suppliers and their associates — and the equally vociferous opposition from established mainframe users and their suppliers.

Companies are based on the principle of *divide et impera*, called delegation. In any organisation, each level acts as an information filter, only passing relevant information to the adjacent levels, and thereby maintaining an acceptable level of information flow across the organisation. The concept of a central company-data base cuts directly across this principle of organisation and this is, perhaps, one reason why data-base systems have not achieved their expected success. In general, the various database activities within a company do not interact strongly with one another, except in terms of summary information at a higher management level. Thus, stock-control packages, sales analyses, ledger accounts and payrolls are relatively free standing. As a consequence, there is no fundamental barrier to the spread of a more independent approach to the processing and storing of information.

It is to be expected that distributed computing will continue to gather momentum, with existing central-computer installations being extended by the connection of separate computers which offload some functions from the central system as it becomes overloaded, and with new systems being built up incrementally on a distributed basis.

The reduced cost of computing also offers the possibility of extending the range of office applications. The major extension is into the area of word processing, as discussed on pp. 133 ff. The relevant question here is whether the new applications will be satisfied by extending the existing data-processing services within a company, or by providing some independent facility. The self interest of data-processing departments and the computer manufacturers will encourage the view that office automation is a natural extension of data processing; however, the disseminated nature of the application and the potentially low cost of relevant semiconductor products, suggests that the data-processing departments will be unable to control the growth of office automation and that it will spring up in an uncoordinated fashion across most companies. From the technical point of view, it is not essential that office automation systems be an integral part of data-processing systems, although the ability to intercommunicate between the two will rapidly become important. The most plausible scenario, discussed on p. 133 is that office automation will mushroom independently and that eventually data-processing systems will be one of the facilities that can be accessed through the automated office, others being an electronic filing system and a central hardcopy print shop to replace the current photocopier machine.

The role of the data-processing manager must be expected to decline. As the computer becomes less esoteric and more accessible to the non-professional user, so the emphasis on specialist computing expertise will decline and the emphasis on general management services will increase. At the same time the reduction in the need for programmers and operating staff, and the physical distribution of equipment, will destroy the rationale which demands that data processing remain a separate function within an organisation.

## Conclusions

*The declining cost of hardware will change the general approach to data processing, leading to the adoption of standardised products which can be used by non-specialist staff, and which are made to match the requirements of specific organisations by their interactive mode of use.*

*The reduced cost of products will enable small companies to adopt data processing, and will remove one of the current disadvantages of the small company as compared to the large.*

*The long-term trend will be away from centralised systems towards a distributed capability which more exactly matches the organisational characteristics of the company. Current data-processing computers may well remain as nodes in the system because of the heavy investments that have already been made.*

*The function of the data-processing manager and his associated staff is likely to decline in importance, and narrow specialist training in these subjects will become increasingly inappropriate.*

## PLANNING

Planning is here defined as the organising and management of resources in a specified way for a given end. Planning is a growth activity, whether it is driven by ideological considerations or the search for economic efficiency. The need for planning is clear at government and at corporation level, and since the achievement of objectives is the basic purpose of any organisation, and has the most leverage on socio-economic issues, it follows that an

126

effective contribution by information technology to this area could have more profound consequences than any other use considered.

## The nature of planning

Just as policy can be defined as a set of objectives plus a programme to meet these objectives, so planning can be defined as the activity necessary to achieve policy. Planning is concerned with the allocation of resources in an organisation and has three main phases:

- the creation of a model which defines how an organisation works and interacts with its environment
- the definition of a plan which determines an expected future state of the model in terms of certain controlling parameters
- plan modification, which requires the monitoring of the actual performance of the organisation, and correlation with the model, in order to adjust the plan to achieve the desired objective and to adjust the model to represent more accurately the behaviour of the organisation.

In many planning activities the model used has been informal or intuitive, partly because there has been a lack of computing capability to implement organisational models and partly because there has been a lack of data to create adequate models.

Information technology can improve the planning process in two ways:

- by providing more powerful and convenient tools for planning
- by improving the quality of data used as a basis for planning.

## THE TOOLS FOR PLANNING

A variety of planning techniques are in use, including input-output analysis, linear and dynamic programming, econometric models, DCF analyses, PERT, production planning and scheduling. Characteristic of all these techniques are the large number of interactions between the variables, leading to the use of matrix techniques, usually based on sparse matrices. The mathematical techniques involved have required the use of the fastest scientific computers with large secondary storage, although currently there is a trend towards the use of dedicated minicomputers to support specific planning facilities. In the future, some form of parallel processing will be appropriate to such models, since they consist essentially of a large number of simultaneous interactions. Since the matrices involved are sparse, arrays of microcomputers may be more effective than simple array computers of the Illiac IV class.

More importantly, the reduced cost and enhanced performance of computing will enable far more interactive modelling and planning, allowing the planner to build up an intuitive understanding of the way in which his system operates, and creating the ability to make better decisions.

## Basic data
The primary failure of most models, particularly macro-economic models, has been the inadequacy of the data on which the model is based. The data available is often only partial, or inaccurate, and is usually considerably out of date. The increased use of data-capture devices in industry and commerce, combined with the growth of electronic information, will in the long term create the opportunity for the collection of much better basic data as a useful by-product of organisational activities. Properly used, such information should greatly improve the planning processes for government and corporations, and will be an important factor in economic competition.

## Conclusions
*Planning will continue to grow in importance as a factor in the operation of government and corporations, and will become an important element in competition between nations and between corporations.*

*Technological developments in computing, particularly in the area of array computers in the medium term and arrays of micro-computers in the long term, will improve current planning capabilities.*

*More importantly, the rapid availability of accurate data will improve the quality of models and the effectiveness of the planning process. The widespread adoption of electronic infor-*

*mation systems will, therefore, have an impact on the overall efficiency of corporations and the economy as a whole.*

## CONTROL OF INDUSTRIAL PROCESSES
The techniques of data processing and planning are common to all corporations, irrespective of sector or function, even though there may be detailed differences in the balance of applications. There would appear to be no similar commonality in the uses of computing to control industrial processes, and as a consequence the rate of development of this type of use has been much slower.

### The current status
The high cost of the computer, together with the associated instrumentation and system design, has restricted its application to a few high-value situations. Typically, it has only been economic to adopt computer control on new plants, and then only on a centralised basis, with one computer system controlling the whole of a large-scale investment. This has meant that the computer system is extremely complex, and it has made the adoption of computer control a high-risk venture.

The use of computers in large-scale continuous plants is well established. Usually computer control has been adopted to overcome some specific problem. For example, in the catalytic cracking of crude, the problem is the variation in the raw materials; in heterogenous phase-catalysed petrochemical reaction, the time constants are too short for effective human intervention; in the production of cement or the continuous casting of steel, high economic return can only be obtained if the plant is very efficient. Typically, in such systems, the control plant represents some 10% of total plant investment, with instrumentation at 7% and computing at 3%.

In the case of batch production, the economic incentive for computer control has been much less. Where computer control has been adopted, it is often for other reasons, for example to obtain the necessary precision in aerospace part machining, or to reduce risk to human life in hazardous environments like nuclear engineering.

Numerically controlled machine tools have not been widely

adopted. In part this has been due to the large increase in productivity possible with NC tools, which can lead to an imbalance in scheduling across a machine shop; and in part it has been due to the high cost of NC tools, which cannot be justified given the small volume and jobbing nature of many small engineering firms in the U.K.

### Future developments
The reduction in cost of computing equipment does little to remove the barriers to the adoption of current forms of computer control. The computer represents only a small part of the required investment, which is dominated by instrumentation and systems-development costs.

What it does do is to open a new approach to industrial automation based on the piecemeal improvement of plant using intelligent instrumentation or local microcomputer control to improve some sub-system. A wide variety of 'smart' instrumentation products incorporating microcomputers is already becoming available, and although most of these instruments are currently oriented to human operation, it is to be expected that an increasing proportion will be designed for autonomous operation with data-communication capability. The addition of a microcomputer enables a variety of improvements to be made to an instrument. For example, with a spectrum analyser, the microcomputer can optimise the resolution of the spectrum, its width and other measures, in order to maximise the spectral information — something previously possible only by human intervention. It can filter the signal to remove noise, and can measure critical parameters, detecting variations outside acceptable limits. Finally, the device can communicate relevant parameters on command, or warning information when limits are exceeded.

To be effective, there is a need for instruments, sensors and controllers to be able to communicate information, either to other devices of the same class, or to computer systems providing some form of integrating control or monitoring. Current interfacing systems, like CAMAC or the ISO standard based on the Hewlett Packard interface, are inadequate, both because of their extremely high cost to interface and interconnect, and because they do not provide any standardisation of the form of communication — are measured quantities represented in binary or decimal, for example? What is required is a general communication system based on a standard information representation (language!), which can be implemented in a variety of physical forms. In particular, it should be possible to support the communication on a multiplex serial channel or on a parallel channel and to have high-noise and low-noise environment versions. Until such standards are evolved, it will be difficult to integrate piecemeal automation in an efficient and convenient fashion.

Piecemeal automation may be expected to stimulate the application of computer control to existing plant. This is particularly relevant to public utilities where the distinction between continuous process and production scheduling is blurred, since demand varies continuously but supply is controlled incrementally. Because of the extent of investment in established technology, and the high cost of providing reserve capacity to cater for peak load, it is probable that the current trend to providing more real-time responsiveness on the supply side, and the balancing of supply-resource allocation by computer will be accompanied by the use of marginal pricing mechanisms to produce more uniform demand and therefore more overall optimal use of resources. In this way, it should be possible to stretch the existing investment to support a much higher level of average activity. These concepts are relevant to utilities like electricity, telecommunications and road networks and could lead to a reduced demand by these utilities for capital investment while the latent capacity is exploited with the aid of microelectronics.

The concept of automation is usually associated with automatic manipulation and assembly techniques. The replacement of the human in such situations requires either a high degree of robotic sophistication, or radical re-design of products and techniques. The development of effective robotic devices is regarded as a hard robotic problem (see pp. 163 ff.) while radical product re-design is not readily assimilated. For these reasons, the introduction of highly-automated assembly techniques is expected to be relatively slow. Again, the piecemeal approach is likely to apply, with certain limited tasks being taken over by computer control, thereby reducing the need for human intervention without eliminating it altogether. Robotic systems like car-body paint sprayers, based on programmed learning by following human

131

actions, will decline in cost and so become more widely used. The development of adaptive operation in such devices, as contrasted to the current mode of operation which is purely repetitive, is unlikely to be feasible at more than a trivial level until the long term.

## Consequences
The progress towards automation has been much slower than expected. There have been various reasons for this:
- the theoretical difficulties associated with developing total control systems and robotic devices
- the disparity betwen applications, which makes generalised techniques inappropriate
- the cost barrier, which has meant that computer control was only applicable to large systems, and therefore had to tackle inherently difficult problems and achieve larger economies
- practical and economic problems of discarding current investments and techniques to take advantage of computer control
- the feared, or actual, social consequences of the introduction of automation, associated with other radical changes into areas where the workforce was often well organised.

The developing technology does not solve these problems. However, by reducing the cost of computer control, it does open the way to piecemeal automation and bring down the level of improvement necessary to make automation worth while. Gradual automation has the potential to be more acceptable to the workforce, since it does not cause the same upsets in job content nor dysfunctions in employment. Even so, the fear of automation is likely to remain, and likely to act as a powerful brake on its introduction.

In the future it is expected that progress towards automation will continue to be slow, even though the development in microelectronics makes it more attractive. Progress is also likely to be very uneven between industries, because of the specialised nature of much automation, and the dependence of many of the barriers on the type of industry concerned. It is to be expected that the direct employment consequences will be small, although the indirect consequences (as improved efficiency reduces the need for new investment) could be larger. Even so, less than 20% are employed in actual manufacturing operations, so that the overall consequences for employment are not very great.

## Conclusions

*Progress towards automation will be slow, and will be based on a piecemeal approach. The emphasis will be on improved utilisation of existing investment rather than on complete replacement. This is particularly relevant to public utilities and could have implications for the associated capital-supply industries.*

*The primary barriers to the application of computer control are not technological. This, together with the application-dependent nature of automation, makes it unlikely that there will be a large return on money invested in promoting technological developments in this area.*

*The lack of adequate communication standards, both at the interface and at the working level, is a serious obstacle to the integration of piecemeal investment in automation. This is an area where investment in development and subsequent standardisation would be valuable.*

## CONSUMER PRODUCTS

The use of microelectronics in consumer products is the most visible example of the potential of microelectronics. Already, advertisements describe the computer clock and the computer-controlled sewing machine — although it may require some stretch of the imagination to describe the microelectronics used as 'a computer'. For the first time, the population as a whole is being exposed to the concepts of information technology; and, indeed, the main consequence of the use of microelectronics in consumer goods may be an increase in public awareness of the technological changes that are occurring.

### Development of the market

The semiconductor industry sees the consumer market as a large

growth area: by displacing existing mechanical or electromechanical control systems, it can obtain access to large new
markets. While this view may be valid in principle, there are
many barriers which will make the penetration of microelectronics
into consumer markets a great deal slower than the semiconductor
industry would like. Principal amongst the barriers are that:

— Microelectronics represents a difficult and alien technology, particularly in current versions of microprocessors.
  Its penetration into consumer markets will be hindered
  by the lack of technical expertise and by a lack of awareness on the part of management. It is natural to expect
  a diffusion of the technology away from existing elec-
  tronics-oriented industries through electromechanical
  industry and finally to the mechanics-based industries.
— The high front-end cost of microelectronic systems
  demand relatively high short-term manufacturing costs
  of microelectronic systems and confines application to
  high-cost products, which can absorb the differential
  over current electromechanical or mechanical alternatives.
— The pattern of introduction of microelectronics is to
  introduce it into top-of-the-range products. This is because
  manufacturers can risk experimentation in the small
  market, whereas they cannot take the risks in their volume
  markets, because in most cases microelectronics offers
  enhanced facilities rather than reduced cost and because
  customers may be expected to be more susceptible to
  the lure of the microcomputer.
— Early application will represent a direct replacement of
  electromechanical or mechanical sub-systems rather than
  thorough-going product re-design, so that full advantage
  is unlikely to be taken of the potential of the microcomputer.
— The introduction of microelectronics requires at least
  some skill changes within a manufacturing operation,
  and perhaps more importantly, skill changes in the
  servicing activity. Such changes represent a barrier for
  both management and workers.

## Mechanical products

Diffusion of microelectronics into mechanical industries may be expected to be slow. Paradoxically, the automotive industry has been one of the first to adopt microcomputers. The reasons are:

- the fact that legislation by the US government on exhaust emission and fuel economy has proved incompatible with the use of conventional technology
- the surprisingly high electronic content of the car, which already has some 2000 components per car representing 5% by value
- the high volume and the capital cost of the car and the industrial vehicle, which makes the front-end cost and short-term manufacturing cost of microelectronic sub-assemblies acceptable.

The automotive industry has emerged as the largest short-to-medium-term user of microcomputers. Already a number of up-market US cars like the Toronado have microprocessor-based ignition-control systems, as part of the learning process for the manufacturers. The US manufacturers are fully committed to microprocessor-controlled ignition, which will be used almost universally in the 1980 models. European and Japanese manufacturers will follow suit.

There are many other points in a car system where the microcomputer has relevance; obvious examples are in brake-control, instrumentation and transmission systems. A less obvious example is the simplification of the electrical wiring in a car. Each application may be expected to be treated as an independent sub-system containing its own microcomputer, and at least in the short to medium term there will be no intercommunication between these sub-systems. The sub-system approach is necessary to provide a coherent maintenance strategy, and it also allows independent development of each application.

Although the US automotive industry has accepted the importance of the microcomputer within the car, the growth of applications will be slow by comparison with semiconductor-industry expectations. The planning and development cycle of the motor car is long, and major sub-assemblies such as engines or transmissions must have a long (twenty years) production life to write

135

off the investment in production plant. Accordingly, applications which involve major changes — such as microcomputer-controlled automatic transmission — will be introduced slowly, even though there may be obvious advantages.

It might be expected that by 1985 the typical American car might contain three microcomputer-based sub-systems, and the leading-edge car five:
 — ignition control
 — partial instrumentation
 — car security
 — brake control
 — diagnostics.

Because of their higher cost and their greater demand for efficiency and safety, it may be expected that many of the innovations will be introduced on commercial vehicles, and only subsequently transferred to private cars.

For the long term, the more important developments are likely to be collision-avoidance systems and in-car traffic-information systems, enabling more efficient use of the road network than can be achieved at present.

The automotive industry is discussed further on p. 182, below.

**Electromechanical products**
Many electromechanical consumer products, particularly in the white-goods area, are potential homes for microcomputers. Under development, and in some cases already on the market, are microprocessor-controlled washing machines, dishwashers, cookers, microwave ovens, food mixers, power tools, sewing machines, etc. In each case the microprocessor is used to replace an existing electromechanical-controlled device.

The advantages of the microprocessor, particularly in the form of the microcomputer, are, in order of increasing importance:
 — enhanced facilities
 — improved reliability
 — easier variability between models, particularly new introductions
 — diagnostics

To set against these advantages, there are often considerable problems because the existing products have been optimised against electromechanical-control technology, and there may be hidden costs or inefficiency in direct replacement by micro-electronics. For example, the inlet actuators on a dishwasher are mains operated, so that it is necessary to introduce a triac control on the output of the microprocessor controller, until such time as low-voltage actuators become available.

In general, electromechanical products have a shorter life cycle than mechanical products, and this, together with the (relatively) greater electronic expertise of the electromechanical companies, should mean that many consumer products with microcomputer control will be introduced in the medium term. These products will use microelectronics as a direct replacement for, or as an extension of, the existing electromechanical-controls system. As such, they are unlikely to require very powerful microelectronic capabaility, or to extend significantly the capability for the user. The most substantial impact may be on the service philosophy, since it should in general be possible to provide specific test facilities for the service engineer, thus greatly simplifying his job.

**Electronic products**
It is in the area of electronic products that microelectronics must be expected to make its major impact. Although not so well publicised as developments in the white-goods market, the use of the microcomputer in products like television and hi-fi is being progressed rapidly, and is likely to create large markets in the short to medium term and to cause great changes in product design. The electronic industries have greater expertise to exploit microelectronics, and the microcomputer is more clearly applicable to their products — both because the technology is common and because the microcomputer is more appropriate for processing information than for trivial control tasks.

In the short term it is the hi-fi market that offers the greatest attraction because its customers are often technology-oriented, and because the data rates are appropriate to current technology. The microcomputer can find application in improved tuning and controls, in fine speed controls on turntables and tape decks, and perhaps more interestingly in conjunction with digital-control signals on cassette tape to allow automatic track-selection and improved noise-suppression facilities. In the medium to long term there is the possibility of using direct digital recording to obtain improved fidelity.

The data rates in television are much higher, so the application of microcomputers in the short to medium term will be restricted to control facilities. These can improve selection and tuning and provide split-channel viewing, ceefax-type facilities and diagnostic capabilities.

## Information products

The final category are those products which, at least potentially, are dominated by their semiconductor content. So far two examples have emerged, the calculator and the electronic watch.

In both cases the pattern of development has been similar, although the struggle in the watch industry is not yet over. There was an existing market based on mechanical or electromechanical products. The potential of microelectronics was identified not by the existing industry, nor by the semiconductor industry, but by new innovative companies. Initially, prices were high, but they declined rapidly under competition, particularly as semiconductor companies entered the market. The original mechanical companies in the market were unable to withstand the competition and withdrew, either directly, or after attempting to acquire semiconductor technology. The innovative companies were also forced to withdraw, leaving the semiconductor companies in command of a market in which prices had fallen to distress levels.

The dynamics of the semiconductor industry suggest that the same pattern of competition can be expected wherever information products that can be implemented directly in semiconductor form can be identified. Looking forward to the future the number of such products can be identified:
— typewriters (see pp. 133 ff.)
— computers (see pp. 122 ff.)
— television (see p. 140) (cameras with optional permanency of image capture are a close product)
— hi-fi (see p. 140)
— books (see p. 140)
It is these industries that are most likely to be impacted as the current techniques are completely replaced by microelectronic

products that can be supplied totally by the semiconductor companies themselves.


## Personal computing

There has been a rapid growth of the 'home-brew' computer market, and a number of forecasts have been made of the development of personal computing, where each individual owns his own computer, comparable in power to the mainframe computer of today, and which he can program to his individual requirements.

There is little doubt that the technology will be able to deliver this capability at reasonable cost and in a convenient portable form (like a notebook) in the long term. The question is whether the average person would have the interest, or the capability, to use such a device. Two alternative views are possible:

— that this new tool will unleash creative potential, enabling users to do things that were previously thought beyond their competence
— that programming and creative problem solving are intrinsically difficult tasks, beyond the capability of all but a few.

There would seem to be no adequate basis for forming a judgement between these alternatives.

A cautious view of the future might be the following:

— In the short and medium term, personal computing remains a hobby market, restricted by the inconvenience of programming and using the equipment. As a market it might be expected to be comparable with other technology-based hobbies like amateur radio.
— In the long term, a personal-computer market develops on the basis of pre-programmed packages which can be marginally tailored by the user to his personal requirements using simplified interactive languages. Such packages would execute on an extended version of the electronic typewriter discussed on pp. 142 ff.
— In the very long term, sufficient capability is developed for the personal computer to 'understand' English text (see pp. 176 ff.), at which time its application can become more general.

On this basis, personal computing is unlikely to emerge as a major market until the long term.

## The long term

The use of microelectronics to replace control systems in current consumer products is unlikely to be a large-scale market for semiconductors in the long term. There are perhaps some 500 million homes in the major industrialised countries: assuming that the microcomputer achieves twice the penetration of the electric motor, and that there is a replacement cycle of five years, this amounts to a total market of 3000M units p.a. At a dollar apiece, this is equivalent to the 1978 semiconductor market in total.

This conclusion is not surprising. Existing mechanical- and electromechanical-control systems are extremely clumsy, and are capable of only limited operation. Their replacement by microcomputers places little demand on the capability of the semiconductor devices. It is only when the current products and applications are transcended by truly innovative applications of microelectronics that substantial demands can be expected. In general, these will involve a high degree of innovation and the creation of new markets, a process that can only be expected to develop gradually. To revert to the electric motor analogy used in Chapter 3, it is noticeable that innovative uses of the electric motor are still being found at a high rate, even though the fractional horse-power motor has been available for fifty years (recent examples of consumer products include the tangential fan heater and the electric toothbrush).

The basic capability of the microcomputer is to process information, and it is to be expected that, at least initially, the innovative markets will be created by replacing existing information products with semiconductor equivalents with the same, or improved, facilities. Using this rationale, it is possible to identify three future markets: television, records and books.

Present technology is inadequate for digital television, but projections of the current rate of development suggest that fully digital television should be feasible in the mid-1980s. Small black-and-white LCD television displays have already been demonstrated, and projections on the performance of the microcomputer suggest that it should have adequate bandwidth to support direct digital processing of television signals at that time. Initially, digital television would offer the possibility of smaller and simpler television sets, and bandwidth compression on television signals. In the longer term, in the form of telespectacles, such systems should allow a similacrum of three-dimensional viewing, at least equivalent in quality to stereophonic sound.

Without compression, an LP record represents about 32 Mb of digital information. Projecting storage costs, silicon digital cassettes should become competitive with records in the late 1980s. The silicon cassette offers a number of advantages, principally a lower-cost transcription device and higher fidelity. The breakthrough of semiconductor technology into a high-volume disposable market of this kind will be an important factor in sustaining the growth of the semiconductor industry.

A book requires considerably less digital information for its representation, some 4 Mb in uncompressed form. The barrier to the use of electronic books is the provision of a reader at a suitably low cost. As discussed on pp. 142 ff., the development of the electronic typewriter should eventually provide a device capable of reading electronic books at a cost acceptable to the consumer. The date at which this might happen is rather speculative, but electronic books should become practicable at much the same time as electronic records.

In the case of both electronic records and electronic books, there is the alternative that the information could be obtained direct over an improved telecommunications network (see pp. 176 ff.), by the equivalent of a library service. While this may be practicable in the very long term, and indeed this would seem to be the natural method for disseminating information, it is unlikely that the coordinated investment necessary to achieve this capability will be made before electronic records and electronic books become feasible, so that they may well be able to establish themselves as an acceptable method for disseminating information prior to the advent of general electronic information services (see pp. 159 ff.).

## Conclusions
*Given many barriers, the growth of microelectronics in consumer products will be slower than expected by the semiconductor*

*industry. Although conventional consumer products represent a large medium-term market for microelectronics, it is unlikely to be a large market in the long term.*

*Because of the relatively low level of demand in the U.K. for advanced consumer products, and the U.K. concentration on down-market products and rather lower level of technical awareness than is shown by some of our industrial competitors, there is a real risk that sectors of the consumer-goods industry will become uncompetitive because they fail to exploit microelectronics. This is discussed further in Chapter 7.*

*Apart from the category of information products, the microelectronic end of most 'smart' consumer goods will be low. In the short to medium term, it is more important, therefore, to ensure exploitation of the technology than to ensure its supply. A particularly serious barrier to the exploitation of microelectronics, although by no means the only one, is the lack of technical awareness and technical expertise, particularly about the microcomputer, in many industries. This could be alleviated by appropriate training courses for management and technologists.*


## OFFICE AUTOMATION

The term 'office automation' is used to cover the various uses of information technology in the office relating to the handling of textual information, as opposed to data processing which handles numerical information. In some cases, the boundary between these two uses may be tenuous in the extreme. At present, the use of office automation is only embryonic, but it is expected to grow rapidly and to be the single most important use of information technology in the long term.

### Principles of text processing

When handling text, the computer has no knowledge of the meaning of the textual information, merely regarding the text as an arbitrary string of characters which can be stored, retrieved or manipulated on the basis of its character pattern. That this is so is readily illustrated by the fact that the same text-handling system works equally well with English or with French texts (providing the operator is English). In this respect, text handling

142

differs considerably from data processing, where the computer has an inbuilt concept of number, enabling it to perform a wider range of transformation on data than it can on textual information.

Although the computer is unable to understand the text, it can still perform a range of useful functions, enabling texts to be stored, transmitted, retrieved or transformed. The transformation can either be under the control of programs which manipulate the text on the basis of the character pattern, for example providing automatic line justification, or under the direct control of a user who can edit the text according to his requirements. It is this latter case which provides the power of a text-handling system, and so, unlike data processing, text handling is essentially an interactive use of computing.

In the long term, it is to be expected that the computer will be programmed to have some direct understanding of the text, providing more powerful and automatic facilities. Because of the complexities of the English language (and the lack of understanding of the workings of natural language) such capabilities are likely to arise by adopting conventions about the form and significance of various parts of a text, so that the computer can 'recognise' an address or the structure of an invoice. Such conventions may be restricted to 'headers' at the start of a text, or may be imbedded within the text itself. The development of these conventions needs careful coordination to ensure that the conventions evolved in different environments are not in mutal conflict. Such conventions should be recognised as another facet in the development of formal language systems, and should be closely integrated with the development of programming languages.

**Current status of word processing**
Current word-processing systems fall into two distinct classes derived from separate ancestors, the automatic-letter writer and the computer-program editor. Most of the available word-processing systems are little more than an electric typewriter with some additional features, the most basic of which is the provision of a memory. The 'memory' typewriter enables the typist to go back and correct/edit selectively, without having to retype the *un*changed pages of text. More sophisticated word-processing systems are similarly based upon the typewriter, but

include higher-capacity memory, based upon magnetic media such as cards, cassette tapes or discs; these enable a large variety of standardised letters or other documents to be automatically typed, requiring only the variable details to be filled in by the typist.

Although such devices are, in computer terms, primitive, they already provide substantial advantages. Much typing is repetitive, and involves only minor alterations of previously typed text. Thus a device that enables text to be typed once, and then to be altered subsequently can, in an appropriate environment, greatly increase the output of a typist — for example in personalised advertising mail, legal work or report typing. Systems of this type have been on the market for some years. Typically the cost has been ten or more times that of an office typewriter and they have not achieved wide penetration except in specialised applications.

The computer-based word-processing systems were derived from situations, such as in programming, where the handling of text has become a substantial part of the activity, and also from the specific text-handling problems involved in applications like the computerisation of newspapers. Computer-based systems tend to offer complex editing facilities, and a variety of other capabilities like text-filing and message-passing between terminals. As such they can be very powerful, but they are often inappropriate for the inexperienced computer user. The capital cost of such systems has been extremely high and, although the word-processing capability may be a useful adjunct, systems have not been purchased just on the basis of their word-processing capability.

The past two or three years have seen the growth of a number of intermediate systems, based on microprocessors or mini-computers. Most of the editing typewriters now incorporate a microprocessor and have enhanced abilities. More interesting is the development of visual-terminal systems. These are usually based on a minicomputer and offer multiple-terminal capability. The computer provides a centralised printing and storage facility, thus allowing the cost of the terminal to be reduced. Using this kind of system, documents can be displayed on the terminal as they are created, and corrections can be made instantaneously. Such systems offer a number of advantages. Obviously, they can perform the function of an automatic-letter writer and also edit

144

reports or similar documents. Also, because they enable better layout and justification, while errors are corrected electronically before the final copy is produced, the quality of the output is greatly improved. This can be valuable both in a prestige situation or where low-quality typing effort is being used. Finally, productivity is improved considerably, both because error correction is fast and because typists work more quickly — particularly near the end of a page — since there is no longer a fear that a page will be wasted due to errors. As yet these systems are expensive, with basic systems costing £10K or more and each terminal £2K-£5K. Even so, there is clear evidence that these products do satisfy a market need and that there will be a substantial demand as the price declines.

There are two distinct directions in which the office automation market could develop; the provision of stand-alone terminals to replace the both ordinary and electronic typewriters, or the extension of data-processing systems to provide word-processing capability.

**The electronic typewriter**
The basic electronic typewriter requires the following: a keyboard, logic to provide the editing functions, a temporary output display, a storage medium, and an output printer. The microcomputer provides the ideal mechanism for such a device, and offers the prospect of low cost, providing the other elements are also inexpensive. There is no problem with the keyboard, and a CRT display can provide an adequate temporary output. Most CRT displays currently have limited capacity (less than 2K characters), but the standard television CRT does have adequate resolution to display a complete A4 page of alpha-numeric text on a 9 × 5 dot matrix. The CRT does however, involve difficult packaging, voltage requirements, etc., and it must be expected that the CRT will rapidly be replaced by a solid-state display, probably based on LCD technology. The storage medium will initially be cassette tape or floppy disc. These have the disadvantage of mechanical technology and intrinsically poor reliability, and, again, are likely to be displaced rapidly by semiconductor storage cassettes. It is both uneconomic and unnecessary to associate a printer with each electronic typewriter, since the typist can have direct

145

electronic display of the page. The use of a shared printer would seem to create no greater difficulties than the current practice of using a shared copier. Depending on duty cycle and volume, various types of computer printer might be used, although for most situations the daisywheel printer should provide adequate capacity combined with mechanical simplicity. Another alternative, which may be expected to emerge, is the use of photocopier mechanisms to produce output direct from electronic representation. In the long term, the requirement for hard-copy output will dwindle as an increasing proportion of users have electronic typewriters and are able to read information transmitted in electronic form.

In the short term, most electronic typewriters will use CRTs and magnetic storage; even so, the price may be expected to decline rapidly to a level where the electronic typewriter becomes price competitive against the more expensive office typewriter. More interesting is the fact that the electronic typewriter has the potential to become a completely semiconductor product in the medium to long term. Such a product would open an entirely new market to the semiconductor companies, and it is to be expected that they will pursue it vigorously. There would seem to be a close parallel between the typewriter and the calculator market, with electronic typewriters also having the potential of a very large home market, provided that prices can be reduced to a sufficiently low level by sales to the business market. Speculation on the long-term cost of an electronic typewriter is difficult, but there would seem to be little reason why, in the long term, it should cost more than the pocket calculator does today.

### The automated office
Typing is only one of the office functions. Other functions of at least equal importance are the filing and cataloguing of information and the dissemination of information both within and outside the office.

The full benefit of the electronic typewriter will only be realised when it is capable of interconnection with other electronic typewriters and also has access to facilities like a central filing system. What is not clear is how this integrated capability will be achieved. Apart from questions of technical feasibility, the question is complicated by the variety of competitors that may be expected to enter this market:

— The mainframe-computer manufacturer must see the office-automation market as a natural extension to data processing — a particularly important extension, since the current market is showing signs of saturation. In order to protect his established market base, he will want to use the new terminals to increase the load on his central system, thereby justifying further upgrades. Because of the limitations on terminal capacity with most mainframes, he might be expected to offer office-automation subsystems that support terminal intercommunication but depend on the mainframe for central storage and extended processing capability.

— The minicomputer manufacturer sees the office-automation market as an opportunity to compete with the mainframe manufacturer. Accordingly he will sell the concept of an entirely separate office-automation system, based upon a central computer supporting terminal intercommunication and providing text-oriented facilities for filing information and input-output. Such systems will also offer interconnection to existing mainframes in order to be able to compete with the mainframe manufacturer.

— The copier manufacturers, and Xerox in particular, are also interested in this market, which in the long term will destroy their current business. Their approach will be similar to that of the minicomputer companies, except that there will be much greater emphasis on the copier and the print room as the focus of of activity.

— The PABX system offers another route into the market. The PABX already provides the method for voice intercommunication within and outside a company, and it might seem natural to extend its function to provide text intercommunication. In practice, the communication characteristics are rather different, with many more lines open for text so that a different kind of switch is required. There is, however, some rationale in attempting to use the telephone connection, since a new physical interconnection within an office could prove expensive. On the whole this does not look like a very good medium-term route, in spite of the IBM 3750 intelligent PABX. In the long term, it may be expected that the PABX systems will be re-designed to use some form of serial-communication loop, and at that time the integration of voice and text would be attractive. It would seem more difficult for existing PABX manufacturers to compete in this market, and they are one of the sectors likely to be threatened by the development of the automated office.

— There are existing companies (Olivetti, for example) in the office market who already have market coverage and have the necessary technical expertise to enter with innovative product concepts. It might be expected that such companies would use a terminal-oriented approach providing a number of functionally differentiated central products including a message-switching system, a central filing system and a central printing facility.

— Finally, miscellaneous new entrants must be expected, together with the existing mechanical-typewriting industry. Because of the scale and the technological requirements, it would seem unlikely that such operations will be successful.

Although major companies like IBM and Xerox are putting a lot of effort into the office-automation market, it is difficult to see them as being successful in the long term. In the short term, such companies may be expected to capture a considerable share of the market by concentrating on sales to large multinational companies, who tend to be more innovative. This total-system approach, however, involves large investments by the users and must be contrasted against the strategy of selling individual low-cost terminals that can be bought speculatively without any major policy decision by the purchaser. There would seem little doubt that this latter route will be preferred, and that it will lead to an explosive growth in the use of electronic typewriters in the medium to long term. For obvious competitive reasons it is the semiconductor manufacturers who will be the main beneficiaries, and it must be expected that they will come to dominate the terminal market. The pattern of competition for the central

148

services is less clear cut. It would seem most likely that a communications systems will develop to which plug-compatible services can be added by alternative suppliers. Such services would include a data-processing facility, a text-filing facility and a hard-copy facility. Essential to such a pattern of development is the creation of intercommunication standards, but it is not clear how these will evolve. The risk is that a variety of intercommunication techniques will be used, and that a *de facto* standard will only evolve slowly, leading to considerable wastage of the investment during the process.

## Electronic mail

The growth of electronic typewriters will create an environment in which electronic mail becomes economically attractive, at least for inter-company communication. It is difficult to see a feasible route whereby electronic mail could develop independently, because of the start-up costs to establish an adequate community of users. The electronic typewriter, however, can be justified from the benefits it creates in stand-alone mode, and its subsequent use for communication between companies will arise as a natural added bonus.

From the point of view of the company with electronic typewriters, electronic mail presents a number of advantages:

— It is already cost competitive with foot mail, and this advantage must be expected to grow with time.
— Where necessary, messages can be transmitted with urgency.
— Electronic mail will integrate more readily with the electronic information-handling mechanisms within the company — e.g., distribution, filing of mail.
— It can eliminate the unnecessary data-prep operation necessary to recapture data-output (e.g., orders, accounts) from computers in other companies.

Electronic mail is already in use, particularly in America, to communicate between geographically separate parts of a company. The extent of this practice is difficult to ascertain, because companies are somewhat fearful of the attitude of the PTTS, nevertheless it does appear to be growing rapidly. Projections for the rate of growth of electronic mail in the U.K. are conjectural. Demand for the facility will depend on the rate at which the automated office becomes a reality, and also on the attitude and pricing policy of the BPO. It is anticipated that there will be informal use of electronic mail in the medium term, and that formal systems will develop in the long term, perhaps before 1990.

The consequences of the use of electronic mail for inter-company communications are substantial. Some 30% of all mail traffic is inter-company. The withdrawal of even a part of that load from the foot-mail system could have a radical effect on postal economics, because the highly-concentrated nature of business mail means that it contributes disproportionately to the profitability of the postal service.

A further 40% of mail traffic flows from business to household, and 15% from household to business. Most predictions of electronic mail look forward to an intermediate stage in which this mail is handed by electronic transmission to a local Post Office for printing and either foot delivery or customer collection. Given the potentially low cost of the electronic typewriter, a more likely development is that direct electronic transmission to the house will be economic within the same time scale.

Obviously, the development of electronic mail has the greatest importance to the future of the BPO, a subject which is considered further on pp. 163 and 228 ff. Another consequence is the reduced attraction of facsimile transmissions. Even given data compression, facsimile requires about ten times the bandwidth of electronic information, making the transmission economics dubious in comparison with postal charges in the short and medium term. By the time that facsimile does become competitive the growth of electronic mail should be in full swing, so it is unlikely that facsimile will ever achieve importance.

At present there are some classes of communication where handwritten or graphical representation is important, and where facsimile might be thought to have a role:

 — Intimate communication. The question is — what premium will people be prepared to pay for the privilege of manuscript communication? Probably it is not very high. Also, in the very long term, it is questionable whether people will need to retain the ability to write in manuscript form.

- Signatures and authentication. A considerable proportion of communication needs authentication. Fully electronic alternatives to manuscript signatures are possible, and are discussed on pp. 154 ff.
- Graphical and picture information. This category is important, and unavoidable. Even here the number of paper originals that require facsimile scanning will decline rapidly as electronic graphical systems and electronic cameras become more widely used.

Thus, even in these cases, facsimile is unlikely to have a long-term role, although it may be required during a transitional phase.

Another aspect of the mail system which is not reproduceable by electronic mail, is the confidentiality provided by enclosure in an envelope. In practice, confidence in this confidentiality represents confidence in the integrity of the postal service, and the foot-mail system does not provide a significantly higher degree of confidentiality than can be achieved electronically. This, however, may not be the view of the potential user. To improve confidence in the security of electronic mail it may be necessary to introduce some degree of encryption (this is discussed further on pp. 154 ff., below).

## The effects on organisation and employment

There are over 1,000,000 secretaries and typists in the U.K. There are a further 750,000 involved in administrative work, and a further 400,000 in management. The advent of the automated office will effect the working lives of all these people, by changing the pattern and efficiency of information flow.

The electronic typewriter increases productivity and could lead to a direct reduction of labour in typing pools. However, a secretary does far more than simply type, and although the electronic typewriter — and, more importantly, other aspects of the automated office — will improve efficiency, this may not lead to a proportional reduction in the level of staffing. In the first place, a secretary is a status symbol which will not be given up readily; secondly, secretaries tend to be distributed throughout an organisation, making it more difficult to eliminate individual posts even if they are under-utilised.

The major change will come only if the electronic typewriter

151

is effective as a working tool for management itself. If this happens, then the use of a secretary as an intermediary becomes unnecessary and large reductions in staffing levels more acceptable. Where the electronic typewriter provides other facilities of direct benefit to the manager — for example, access to personalised data-procesing capability — this barrier may be broken down raidly. Otherwise, the skills and status afforded by the secretary will keep the electronic typewriter firmly in the outer office.

The introduction of direct-speech input would change this situation completely. With direct input, the secretary not only becomes unnecessary, but is a direct obstacle to the effective use of the electronic typewriter. Direct-speech input is, therefore, expected to have a far more radical effect on the pattern of office work and the level of employment of secretaries than the first stage of the automated office. For reasons which are discussed below, on pp. 176 ff., direct-voice input is regarded as a difficult problem and, in spite of repeated predictions and announcements to the contrary, is not expected to become practical in the short or medium term.

Office automation must be expected to have a radical impact on the efficiency of information flow within a company. This in turn may be expected to have an effect on the overall efficiency of the company, particularly in the way it utilises resources, and hence on its competitiveness. Much of management activity is concerned not with decision making, but with the collection and distribution of information. The development of the automated office may, therefore, also be expected to affect the functions of management and indirectly provide a substantial improvement in the manager's efficiency. Thus, office automation may be expected to reduce the number of managers required to operate a company (and thereby to generate a further reduction in secretarial requirements).

Finally, there is the management-consultancy aspect. Much of the improvement in efficiency arising from the introduction of data processing came not from the computer itself, but from the opportunity to examine procedures critically and to change them when data processing was introduced. Much the same pattern may be expected with the introduction of office automation.

The introduction of office automation may lead to serious labour displacement in secretarial, typing, administrative and management staff, and hence to a complex set of social changes. With office automation, the sector of employment under threat is not, as in other cases of automation (data processing, factories, etc.) the general workforce, but the administrative force, with the emphasis predominantly on female workers; so any displacement would have major social consequences for female independence and family income. At present there is high job mobility and a low level of unionisation amongst workers of this kind so, initially at least, the work force would be unable to resist pressure. Management has shown a similar pattern of fragmentation and inability to exert any influence although, in this case, union organisation has been fairly good. Under the threat of unemployment, both sectors may be expected to become more organised and militant. The consequences could be far reaching, creating the opportunity for greater strife between the main sectors of the community.

The adoption of office automation could have a further social consequence. Office workers are highly concentrated into the civil service, local government and the major towns. Thus any labour displacement will be distributed unevenly not only across the population but also geographically, and between the public and private sectors.

Office automation is seen as having far greater social and economic impact than any other aspect of the information revolution, and must be the primary problem to be addressed by any government policy. It is not the worker on the shop floor who has the most to fear from the computer, but the office worker and his manager. The impact of information technology, even in a major manufacturing company like British Leyland, will be greater on the office staff than on the assembly line (in fact, British Leyland employs more men and women in office staff and management than in direct manufacture). Not only are the number of employees involved greater but a greater gain in productivity is almost certainly possible by improving the efficiency of the office rather than the efficiency of the production line.

## Conclusions

*Office automation is expected to be the main growth area for information technology in the medium to long term. By its nature, it requires a different approach to conventional data processing. The existing computer companies will not necessarily dominate the new market.*

*The key to rapid expansion of office automation is the development of an all-semiconductor electronic typewriter. This is feasible in the short to medium term, the only outstanding problem being an adequate semiconductor display.*

*For the long term the key technical development will be voice input, which could have a radical effect on office organisation.*

*There is a need to develop adequate standards for inter-communication within the office environment. Initially these standards will concern the interconnection of terminal devices and the routing of messages. However, to provide computer-to-computer communication between companies it will be necessary to introduce a high degree of standardisation into items like orders and accounts. This needs to be seen not as the introduction of format or protocol but as part of the process of standardising language.*

*The introduction of electronic typewriters will make electronic mail for business activities feasible. The attitude of the BPO towards this development will have a profound effect on the way such communications develop, and possibly on the long-term efficiency of industry.*

*Office automation has the potential for large-scale economic and social consequences. The nature and extent of these consequences should be the subject of further study.*

## ELECTRONIC FUNDS TRANSFER

Electronic Funds Transfer (EFT) refers to a class of systems in which the financial transaction is *implemented* electronically. This is in contrast to a point-of-scale (POS) system, where a financial transaction is *reported* electronically.

Currently, financial transactions are of three types:
- cash, where the transaction is direct in terms of the tokens exchanged
- cheque, where the transaction is authenticated by signature and is carried out within a secure environment, the bank

— credit card, where the transaction is authenticated by signature and is carried out within a secure environment, the bank.

EFT itself already provides a fourth category of substantial importance, in that it is used within the banking system itself for inter-bank transactions.

## Alternative types of system

The key features of an acceptable EFT system are: that it should provide a method of authentication, that the total system should have a high degree of availability and security and that the cost of the transaction should be economic. Three different approaches appear possible:

— Direct on-line transaction. In such a system the transaction would be entered at a local terminal, for example in a shop, and transmitted to a remote computer for authentication and action. Such a system is impractical with the current telephone network, since it would require one telephone line to the computer for each shop. Thus, the approach would either require that the banks develop a private network for EFT, which in time would become very large, or that the BPO provide a message-multiplexing service (e.g., packet switching) with the necessary integrity.

— Off-line transaction. In an off-line system, the transaction would be entered at a terminal, which would have the capability for authenticating and completing the transaction. The terminals would be polled by the bank system periodically to collect completed transactions. Such a system would be practical with the current telecommunications network, but it requires that the terminal should have access to adequate information (e.g., the customer balance) to be able to validate the transaction.

— Decentralised transaction. In this system, transactions are not recorded centrally. Provision is made in the terminal for credit to be withdrawn from the terminal in a manner analogous to entry. This gives a direct analogue with money, the terminal acting like a till.

Of these three systems the decentralised transaction is regarded

155

as the least plausible, even though it might be more acceptable to a considerable portion of the public (because no central record is created of the financial actions of the individual, there is no risk of invasion of 'privacy' by Big Brother, the Tax Man, etc.). The system is implausible because there is no direct financial incentive for its products. The direct on-line transaction is the system most often thought of in the context of EFT, but this system places severe demands on the telecommunication network and on the central bank system. So it is the off-line system which is considered to offer the more attractive approach, at least initially.

In the off-line system the terminal must be able to verify the transaction. This means that the customer must provide identification and evidence of adequate credit. This requires some form of erodable money card, and the most satisfactory implementation would be to use a microcomputer mounted on a plastic card similar to the present-day credit card. This should hold details of past transactions, credit level and customer-identification information, part of which could be personalised by the user to ensure complete protection. To perform a transaction, the terminal would check the credit level, verify the authority of the customer to use the card and finally perform the transaction, storing details on both the terminal and the card. The terminal would accumulate transactions and transmit these to the bank system on request. Typically the terminal would be polled at night by the bank system, thus enabling the transaction load on the system to be controlled and putting the night-time spare capacity of the telephone network to good use.

A system of this kind has a number of intrinsic security characteristics. Firstly, the personalisation of the money card would protect against card theft, and against embezzlement within the system. Secondly, if the bank system reads past transactions when the money card is recharged, this creates a double-entry system which can protect against fraud by both the customer and the vendor. Finally, by limiting the credit accessible through the card, the customer is protected against excessive loss if the card is mislaid.

The main problem with this, or any other EFT system, is to provide an adequate level of security to ensure that electronic money cannot be forged or tapped. To some extent this has been

156

achieved by the off-line system, which imposes a double accounting check; but this is hardly likely to be acceptable on its own. The concept of making the system physically secure must be rejected as impractical for the telecommunication parts (although the use of the public network does make the system more secure than the use of the private nework), and uneconomic for the terminals. As it is, cryptographic techniques have been developed that can provide the necessary security. These trapdoor cyphers are based on special numerical transformation which can be performed efficiently in both forward (encoding) and inverse (decoding) modes, but which have the characteristic that even though the decoding algorithm is known, it is computationally infeasible (i.e., would take too long, except by chance) to calculate the encoding algorithm. Using such a cypher, it is possible to transmit unforgeable messages. The sender sends both the encoded message and the decoding algorithm so that the receiver can read the message, even though he does not have enough information to be able to create the message. However, it is necessary to add to this system to prevent forging by the recipient sending on the same unforgeable message more than once. This can be done by the sender adding a serial number to the message before encoding to ensure that all valid messages are distinct. A further refinement of this encryption technique will allow messages to be transmitted in cyphered form in such a way that they cannot be read by a third party. Such a refinement is not relevant to EFT, although it could be the basis for a secure electronic-mail system with authorisation (see above, pp. 142. ff.).

## Feasibility of EFT

Although EFT is clearly technologically possible, and probably economic in the medium term, it does not follow that it will happen. The creation of an EFT system will require a large initial investment and it must provide an adequate return to whoever makes the investment. With present money systems the costs are distributed widely. The government bears the cost of maintaining the currency; the customers and vendors bear the cost of currency transactions in terms of time and credit; banks and customers bear the cost of cheque transaction; and banks and vendors bear the cost of credit-card transactions. Thus, while an EFT system might produce an overall cost benefit, the benefit to individual parties may not justify them making an investment.

It does seem, however, that there are a number of factors which, taken together, could encourage the banks to invest in EFT in the medium term:

— The present cheque system is reaching the limits of its capacity and the cost of transacting a cheque is unacceptably high.
— The credit-card system has been unsucessful, in that most users are in the AB class and use it as a money card. The banks would like to integrate the credit-card system with the cheque system, but cannot do so without replacing their present computer systems.
— An EFT system would attract new customers to the banks.
— By collecting EFT transactions the banks will be in a position to offer new services to their customers, including personal accounting for individuals and small traders.

With a carefully designed system, it should be possible to keep the initial investment at a reasonable level. The terminal cost can be low because the system can be totally electronic and does not require mechanical protection for security — and its cost could be borne by the vendor. The money-card costs, although low per unit (less than £5), would be high in aggregate — but could be borne by the customer. The bank system could be based on local computers, rather than on a centralised system, the number of local computers being increased as the load builds up on the system. Finally, the usage for the system could be controlled by limiting the minimum value of a transaction, gradually reducing this limit as costs were reduced and additional capacity became available, until ultimately all cash transactions were eliminated.

### Conclusions

*Technologically, EFT will become feasible in the medium term. Because of the new market opportunities it offers and the limitations of existing transaction system, the banks will have an incentive to pursue EFT.*

*EFT will raise in a very clear form the question of privacy, since the thoroughgoing adoption of electronic money will enable* a *full record of an individual's financial activities to be maintained, thus allowing the possibility of surveillance or of more rigorous and innovative taxation systems.*

*The U.K., because of the concentration of joint stock banks and because of its size and population density, is well placed to lead in EFT, as it has done previously in other aspects of the banking system. Properly approached, this could create an opportunity for the U.K. information-technology industry both to satisfy the U.K. requirement and to get a leading position in world markets.*

## INFORMATION SERVICES

The information occupations in the economy have grown rapidly during the past 100 years, from involving less than 10% of the working population to involving over 50%. This includes all those involved with production, processing, distribution and use of information. Associated with this growth there has been a rapid rise since the Second World War in the supply of information as an economic activity — ranging from credit ratings to specialist consultancy. The convergence of computing and communications offers the possibility of powerful new methods for the supply of information, and this is the area where information technology is likely to have the most structural impact in the long and very long term.

### Range of services

The range of information sevices that are amenable to electronic information technology is extensive. Excluding the point-to-point-communications service discussed on pp. 163 ff., in the section on telecommunications, there are a variety of broadcast or multiple-access services that can be conceived:

- The earliest services will be those which provide rapidly changing information of considerable economic value to the businessman. Some examples, like airline- and hotel-reservation systems, stock-exchange transactions and international exchange rates, are already available.
- The second class may be expected to be types of transient information with fewer barriers. Examples are encyclo-

159

paedias, newspapers, shopping services, consumer advice and demand education.
- The third class is for information which would be physically expensive to access and transport by alternative methods. The prime examples are books and current library services.

The total volume of potential transactions is large — several times greater, in fact, than the current usage of the Public Switched Telephone Network (PSTN).

## Pattern of delivery

Three types of electronic information service are already in at least prototype operation:
- computer-based services (like reservations): provided by independent computer systems and accessed through the PSTN
- Viewdata: the BPO service providing selective access to information using a standard television set and adaptor equipment to connect this into the PSTN
- Ceefax: a television service, with information transmitted by radio and accessed by a modified television set.

These systems must be regarded as prototypes for future information-delivery systems rather than as a long-term solution. There are many obvious problems:
- Computer-based services are in competition with Viewdata and there is the risk that the BPO will at some time price them out of the market.
- The Viewdata system is extremely constrained because the access control system has been designed around near obsolete technology. The use of a television display is a short-term expedient. As such it has disadvantages; the television is not physically located convenient to the telephone in most homes (75%); the TV set already has high utilisation for other purposes; and it is in the wrong social environment (recreation) for many Viewdata purposes.
- Viewdata also raises important questions about the future role of the BPO: to what extent should it be allowed to become a monopoly supplier of information services?

Alternatively, should the supplier of information services be the responsibility of a new statutory body? When Viewdata is upgraded, a special terminal will be required. Should the BPO have monopoly rights on its supply, as it does with the telephone? Should the BPO be allowed to take a profit from the services provided, over and above the profit on the data transmission itself? Should the BPO be able to control the nature and quality of the services provided?

— Similar questions apply with the Ceefax service, although the more limited range of information that can be provided, and the current editorial role of the BBC means that the BBC is not adopting a substantially new public role as is the BPO with Viewdata. More relevant to Ceefax is the question of how this service should be charged, and, if it is charged selectively, whether or not this charging should be extended to other television services (a micro-computer television meter).

Ceefax provides a selective-broadcast system. Viewdata provides a selective-access system. Any ultimate distribution system needs to provide both these capabilities, perhaps with intermediate levels of distribution and charging related to the number of people accessing the information. The ultimate system must also provide a more controlled method of charging, both for the transportation of the information and for the right to access and use this information. Thus information services will require a new form of distribution network which should properly arise as the future development of the PSTN (see below, pp. 163 ff).

The aspects of greatest importance for the distribution of information are:

— the availability of an adequate local-distribution network
— the provision of flexible charging services
— the provision of suitable access terminals.

## Social consequences

The BPO and other PTT suppliers of telecommunications capability are necessarily highly regulated, and are not noteworthy for quick decision making, responsiveness or innovation. There are good reasons why this should be so, but the time constants for the BPO appear out of balance with the rapidly evolving nature of telecommunications potential. There is a strong possibility that pressure will develop from independent companies like IBM or Xerox to enter the market as purveyors of electronic information, and increasing legal conflict must be expected between the statutory bodies and the large corporation who see an opportunity or are threatened by the potential development.

The dissemination of information electronically could affect a variety of industrial sectors, principally forestry, paper making, newspapers, books, advertising and retailing. The pattern of development will be critically dependent on the kind of telecommunications capability available. It might be expected that limited interactive services like shopping guides, consumer advice, news, entertainment and education on demand may well be easier to establish and therefore weaken demand for fully interactive services such as remote shopping, banking services or fully programmed education services. Given the powerful economic force exerted by the semiconductor companies as they look for new outlets and new opportunities for information services, the outcome may not be a 'fair' competition, with the services that provide the best cost/benefit eventually predominating, unless there is regulatory action of an unprecedented kind.

The dissemination of information (e.g., newspapers, books) is important economically, but so far it is operated largely on a national basis. The use of electronic technology will reduce the cost of dissemination, thereby concentrating the cost on the collection or creation of information. As a result, there will be increased internationalisation of businesses operating in the information service market, as they try to obtain larger markets for their wares. The other consequence is the question of custom duty on services transmitted internationally. For example, a computer which is imported attracts duty; but access to the capability of that computer across an international boundary does not currently incur any customs duty.

The dissemination of information is important politically. The development of information services is a matter of the greatest importance in terms of regulation, control and freedom. Will the social bounds of control on Viewdata data bases be as satisfactory as even that of Fleet Street? Should the BPO have any role as an editor or censor?

## Conclusions

*The development of information services will have a profound impact in the long and very long term.*

*There is a need to plan the development of the telecommunications network to take account of the growth of information services and to provide the necessary characteristics within the communications system.*

*There is a need to re-evaluate the role of the BPO in the provision of information services. Should it act just as a carrier? Does it have an editorial responsibility, or right? Should it be allowed to take profit on the information conveyed rather than on its conveyance?*


# TELECOMMUNICATIONS

Communications may be expected to play an increasingly central role in information technology, and the pattern of development of appropriate services will be a key factor in determining the rate at which information technology is exploited.

## The current situation

The BPO is a public corporation operating under an independent Board, which reports to the Secretary of State for Industry. Its authority is defined by the Post Office Act of 1969, which gives it considerable monopoly powers over the transmission of information. The Act was drafted before the consequences of micro-electronic information technology were apparent, and many of its aspects may be inappropriate to the changing function of the communications system.

The telecommunications system has developed gradually throughout this century and now represents a very large investment oriented towards voice communication using electro-mechanical technology. The BPO has enjoyed a well-defined, but growing market, and has been able to structure itself in a highly functional way to increase its efficiency. This has led to a very inflexible structure which is not responsive to the changing technology.

Data communications, although growing, represent only a small part of telecommunications, currently less than 3% by value. Most data communication is carried through the standard telecommunications network, using a modulated voice signal to carry the digital information. The provision of automatic dialling and receiving equipment means that computers can now communicate directly through the Public Switch Telephone Network (PSTN) without manual intervention. Most digital communication is, however, carried on leased lines which can be selected for quality and enable a higher transmission rate. The maximum data rate achievable on the PSTN is 2400 bps.In general, this limitation is caused by the frequency-multiplexing strategies used on the trunk line and not by the capacity of the local connections. Using private circuits, data rates up to 48 Kbps are possible. The BPO controls the data rates that can be used, and places a constraint on the use of data transmission by acting as a monopoly supplier for the modem equipment necessary to connect to the network. There is no standard basis for charging for the use of private circuits for data and the BPO has recently considered placing a tariff on leased lines on a traffic basis rather than at a fixed cost. This makes the economics and planning of private networks uncertain.

The lack of a generally available data-communications service has forced many large organisations to develop their own private data networks using leased lines. These have proliferated in a way which makes intercommunication between them difficult. The future of such networks must be uncertain, because if and when the BPO does introduce a more general system, it will have an incentive to phase out the private networks in order to provide a viable loading on its own system at the earliest possible date.

Following considerable pressure from the computing community, the BPO introduced an experimental packet switch system (EPSS) in 1975, specially designed for data communication between computers. This was set up to evaluate packet switching as a technique, and to enable the BPO to investigate operational characteristics. It was also expected to provide an indication of demand and a definition of equipment practice as a basis for international standardisation.

EPSS provides a switched data network, using selected voice circuits and special computer-based digital switches. Users connect to the network via leased lines, at rates up to 48 Kbps. In the experimental system, packet-switch exchanges are situated in London, Manchester, Glasgow, with interconnection by 48 Kbps links. By comparison with the voice system, EPSS provides a much lower error rate, the ability to re-route messages in the case of failure, much faster call connection and the ability to interconnect terminal devices operating at different data rates.

On the basis of this experiment the BPO is expected to set up a permanent packet-switch system (PSS) using revised inter-communication protocols based on the new international X 25 standard, with a total of nine exchanges. This could come into operation by 1980, using foreign equipment, but the real demand for this service is not established.

**Planned development of the PSTN**
The growth of voice traffic is unlikely to be very rapid, and should be predictable on the basis of past experience. There is no evidence that the volume is highly sensitive to price, so the pattern is unlikely to be affected substantially by the technological development. Indeed, analysis of the operating cost of the BPO suggests that improved technology is unlikely to reduce costs, although it may prevent them from escalating. The growth of traffic. currently about 7% p.a. can come both from increased use of the PSTN and by extending its coverage to a larger pro-portion of the population.

The BPO is in the process of moving from the use of analog transmission for long-distance signalling to the use of digital transmission in the form of pulse-coded modulation (PCM). This enables the quality of the signal to be maintained more readily through multiple relays, and allows the use of more powerful digital switching techniques. The development of high-bandwidth transmission systems, particularly using optical techniques, means that the trunk part represents a declining proportion of the total investment in the PSTN. Associated with the use of PCM is the introduction of new trunk exchanges based on System X technology, which will provide computer-controlled switching of digital circuits, with a basic capacity of 64 Kbps representing one voice channel.

It is also planned to develop the System X technology to provide local exchanges. This would extend the PCM network, so that only the individual connections between the subscriber and the local exchange remained in analog form, conversion from analog to digital occurring within the local exchange on each subscriber line before switching. It is this area, and the individual subscriber line in particular, that represents the greater proportion of the total investment in the PSTN. At present, only 60% of households in the U.K. are connected to the telephone system, so that any extension to more general coverage would require large-scale investment.

The provision of PCM and electronic digital exchanges should greatly enhance the reliability of the PSTN and enable a reduction in the large maintenance force. From the point of view of the subscriber these developments have little impact except to maintain or improve the quality of service. It is possible to envisage an increased range of basic services, including re-routing, call-back-when-free, etc. Most of these services will depend on the caller's number being accessible through the system and so will require the introduction of local digital exchanges, or, alternatively, the provision of more sophisticated telephones.

### Satellite communications
A satellite-based communications system offers a radically different approach to the present ground-based system in terms of technology. However, in terms of facilities to the user, there is no reason why the two systems should provide differing services. The primary difference is the increased end-to-end delay in a satellite system, which may be as much as half a second. For the user, such a delay is quite acceptable for voice communication; and it is unlikely to be important in most data communications, which may be even more tolerant of end-to-end delay. The main consequence of the idle time will be to encourage more complex switching strategies within the communication system itself.

Promotion of competition within the US by the FCC has led to the viewing satellite-based systems as an alternative to ground-based communication, with the possibility of direct competition.

It would be far more useful to see satellite communications as an extension to ground-based communication systems, providing a more cost-effective solution in certain situations (for example, in sparsely populated areas).

The development by IBM of satellite-based communications through its subsidiary SBS, could have a profound impact on computing and communications, particularly since after 1980 IBM becomes free of legal restraint and able to re-enter the service-bureau market. As indicated in Chapter 5, IBM has the financial resources to build an effective world-wide communications system, and to offer a range of communications facilities, including electronic mail. One possible corporate objective would be for IBM to become the world's communications facility. In principle, the BPO and the other PTTs could prevent IBM from offering services locally, since they (the Home Office in the case of the U.K.) control the ability to receive broadcasts. In practice, if IBM had made the necessary investment and developed a range of attractive services, strong commercial pressure might develop which could brush aside the *de jure* situation, with the PTTs being unable to offer comparable services or to make the investment necessary to provide these services in the future.

### Development of data communications

The relationship between computing and telecommunications has been obstructed by two unfortunate problems. The first is the anti-trust legislation in the US, which prevented Bell from developing computers and IBM from participating in communications, thereby creating an artificial barrier which it has taken many years to break down. The second is the active advocacy by the computing community of packet switching as the most appropriate method for data communications. This has meant that attention has been concentrated on one technique for providing data-communication services rather than on the type of service that is required. Packet switching, which to the telecommunications engineer is a form of statistical switching in both time and space domains, is a technique for transmitting information through a network. There may indeed by a role for packet switching both for voice communication and for data communication, but it needs to be considered against alternative techniques to identify the most cost-effictive solution using the new digital technology.

There is a risk in extrapolating the demand for data communications and the type of services required from past experience. Much of the early requirement for data communications was based on the economics of providing centralising processing facilities, leading to the concept of multi-programmed computers accessed by remote terminals or remote job entry terminals. The changing relative cost of processing and tele communications will make this mode of use increasingly unattractive. The second main usage has been computer to computer, to implement private networks. Again, this mode of usage may disappear with the provision of a Public Switch Data Network (PSDN).

In the future the pattern of data communication is likely to be quite different. The main loading may be expected to be generated by electronic mail, electronic funds transfer and various forms of remote telemetry. In general these services will be person to person, mediated by computer, or person to computer. The dominant data rates will be the rates at which the human can generate information (c. 80 bps), read information (c. 1 Kbps) or scan information (c. 8 Kbps). For psychological and economic reasons it may be desirable to transfer information in bursts, leading to somewhat higher peak data rates. For individual users, voice-bandwidth (PCM) lines (64 Kbps) will provide more than adequate data capacity, and indeed, it should be economic to multiplex many data communication paths on to a single voice path.

Data communication differs from voice communication in certain essential features:

- There is no characteristic bandwidth. The bandwidth required may be expected to vary from station to station, and to vary from time to time at a particular station.
- Data communication has less inherent redundancy than voice. It can therefore, transmit information at a greater effective rate providing that the transmission system has an adequately low error rate.
- Many of the potential uses of a PSDN, like electronic mail, do not require instantaneous station-to-station

168

transmission, or bi-directional communication.

The key to the provision of an adequate PSDN is seen not in terms of the requirements on the trunk communication system, but in the provision of an adequate local communication distribution capability. Because of its characteristics, the most appropriate form for local communications would be a high-speed multiplexed communications line, from which individual stations could tap bandwidth capacity as required.

Attention needs to be concentrated on the fundamental data-communication facilities that are required, rather than on superficial services like Viewdata. At this level, it would seem that the PSDN needs to provide a facility, or a range of facilities, allowing the user to choose the following for individual calls:

— acceptable error rate
— maximum delivery delay.

The other possible dimension is security, but it would seem better to achieve this when required by external encryption.

It appears quite impracticable to predict the long-term volume of data communications, or the pattern of growth. These will be determined by a variety of extraneous factors, which are discussed at various points in this study, and also by the availability of adequate communications facilities to generate the demand. It cannot be stressed too strongly that we shall see the growth of data communications only if the BPO initiates the process by providing cost-effective services, with an adequate guarantee of long-term continuity.

In the medium to long term, the volume of data relative to voice traffice (measured in bits) is unlikely to be high. An educated guess would suggest that data will represent 10-30% of total traffic in the long term, rising to 80 or 90% in the very long term. Thus data-communications traffic is seen as making a useful contribution to the growth of telecommunications as a whole, rather than as generating a new situation. There is one caveat to this view — the possible development of station-to-station picture transmission, for example to enable selective television, picture phone, or teletransportation. Picture transmission requires a bandwidth of one to two orders of magnitude greater than voice transmission, and the development of such services would have radical consequences for the telecommunications system. The

main impact of information technology on the PTTs is seen as:
- changing the technology used to provide communications services
- increasing the importance of telecommunications as an essential service.

Apart from this, the key question is whether the PSTN and a future PSDN should be seen as two separate networks, or as a single network. The arguments in favour of a single network are thought to be compelling:
- There is no serious incompatibility in requirements between the two networks, and they can both be implemented by the same technology.
- The use of a common network will make the economics of the PSDN more acceptable during its early growth.
- It will be difficult to forecast the relative growth of the two types of service, and a single network will provide greater flexibility.
- In the long term, it is expected that most users will require access to both services.
- The PSDN is subject to a variable loading, with very little demand on its capacity outside working hours. By contrast, many of the PSDN services, like electronic mail and electronic newspapers, can operate overnight, thereby providing better utilisation of the total system.

The consequence of this view is that forward planning of the PSTN should take into account the future requirements for a PSDN.

### The role of the BPO

The BPO is seen as having a central role in information technology — as a user, as a supplier and as an organisation likely to be impacted by the changing technology.

The BPO investment programme runs at some £600M p.a., the major part of which will be in the area of information technology. That being said, the BPO will be the largest single purchaser of the technology in the U.K., and could, by the exercise of a positive investment programme, play an important part in the promotion of the technology.

However, the purchasing record of the BPO is extremely poor. Its demand for specialised specifications has made it difficult

170

for the U.K. telecommunications industry to compete outside the U.K., and an irregular pattern of purchasing has led to inefficiency. The seriousness of the situation can be measured by the falling balance of trade in telecommunications, which has moved from positive to negative continuously over the last two decades.

Equally seriously, the BPO has shown little appreciation of the consequences of programmed digital electronics. Its decision in the late 1960s not to pursue stored-programmed digital technology has put the supply industry into an extremely weak competitive position for the future, to a level where the purchase of foreign equipment for stored-program and packet-switch exchanges is under consideration.

The late decision of the BPO to invest in the System X technology must also be regarded with concern. There is little evidence that the BPO have the necessary expertise or capability to carry through a development of this kind successfully, or indeed that System X has been based on a good appreciation of technological development. The possibility of failure with System X must be regarded with the greatest disquiet. Not only would it destroy the U.K. telecommunications industry, and with it probably the remainder of the information-technology industry, it would also prevent the effective exploitation of information technology within the U.K. because the necessary communication facilities would not become available.

Relevant to the CSERB, the BPO has a research and development staff of over 1500. With the convergence of technology this effort can no longer be considered as separate from other R & D activities in computing, and some means needs to be found to achieve coordination or better integration.

Regrettably, from the point of view of information technology, the record of the BPO is again extremely poor. Much of the early development of packet-switching techniques was carried out at the National Physical Laboratory (NPL), which achieved a world lead in the subject. This lead was not exploited by the BPO or the government and has not been taken up by the BPO research organisation. Currently, within the U.K., there is very little on-going research in the critical areas of digital data-switching techniques, and the implementation of an effective local-distribution network. What little research there is occurs at the

NPL and in an uncoordinated way at London University, supported by the Science Research Council (SRC).

The distinction between digital processing and digital communications is becoming theological rather than practical, and this raises fundamental questions about the role of the BPO. There are three main questions:

— Should competitive suppliers of data communications be allowed?

— Should the BPO be permitted to extend the range of information services it provides?

— Should the BPO be more positively integrated with other aspects of information-technology policy and decision making?

In the US, the Federal Communication Commission (FCC) has overturned the monopoly position of AT & T and is permitting a considerable degree of competition in the provision of communication services. It is not clear that there is any justification for a similar approach in the U.K. The provision of telecommunication services seems to satisfy the criteria for a natural monopoly.

— There are identifiable natural economies of scale.

— There are direct benefits to the user in a monopoly service.

The difficulties with the monopoly situation are those of ensuring that the BPO provides a cost-effective service and that it provides new services to the benefit of the public, rather than for its internal interest. While it may have been adequate in a static situation, the current structure of the BPO does not meet these requirements; the control and organisation of the BPO needs revision of its continued operation as a statutory monopoly is to be acceptable. In particular, the decision about the overall scope of the telecommunications system and the services provided need to be taken by people not directly concerned with the immediate operating profitability of the corporation. Within the wording of the Post Office Act, the nature of the BPO monopoly powers is somewhat ambiguous. The BPO has interpreted these powers on a wide basis, and this interpretation has not been challenged in the courts. The BPO claims the right of supply of attachments to the PSTN and may also be considering a monopoly position on the provision of some value-added services using the PSTN. In neither area does there appear to be a natural monopoly, nor any justification for maintaining a statutory monopoly. Any attempt by the BPO to achieve a preferred position operates directly against the interests of encouraging an active information-processing industry, and it must discourage independent suppliers from developing innovative products (whether attachments or services) either because of direct prohibition (e.g., in the case of modems) or because of the fear of future BPO action. The position of the BPO is understandable, given the directive from the government that it is to maintain an overall profitable position; but a better objective would be to maintain individual services like the PSTN at a given level of efficiency, and not to achieve profitability by cross charging from other activities.

There must be concern that the BPO's interpretation of its monopoly position and its requirement for profitability will inhibit the development of some part of the information-technology industry. Of even greater concern must be the possibility that the pursuit of commercial objectives by the BPO will lead it into the role of Big Brother forecast by George Orwell in the book *Nineteen Eighty Four*.

The development of data communication and the wide range of services and opportunites it creates renders the present Post Office Act innapropriate, and there is a need to redefine the responsibility of the BPO in the new context. Such a definition might be based on three principles:

— The BPO should have a monopoly right in the transmission of information between remote points.
— The BPO should have no rights to control the form, significance or usage of such information.
— The BPO should be excluded from the provision of equipment or services using the PSTN. If it is thought advisable that there should be publicly owned corporations in this area, these should be operated and accounted independently from the BPO.

The tariffing of BPO services is almost as important as their availability, since that will determine their viability. In the past, the BPO has been able to operate a cavalier tariffing policy, because most users have no direct investment in the system and the tariffing only affected their revenue decisions. Now, with the emergence of a large computer network, whose raison d'être is

173

to minimise the cost of utilisation of the PSTN, the user's investment decisions are far more dependent on the current and future tariffing policy of the BPO. In order to provide a rational basis for investment decisions, it is necessary that the BPO adopt a more transparent tariffing policy, and that it should undertake to maintain the *principles* of the policy for a sufficient period to justify investment decisions. Such a tariffing policy must have several components; of particular importance is the relative pricing of data against voice services. A desirable set of charging principles might be:

— total tariffing to cover operating costs, interest and depreciation (based on replacement using latest technology)

— call charging on the basis of per-bit capacity used, independent of bandwidth or usage (in particular, whether voice or data)

— call charging independent of distance — application of the 'Rowland Hill' principle

— variable tariffing to balance the diurnal load on the system.

The final area where the BPO interacts with information technology is in terms of employment. Currently, the BPO employs 410,000 people, of whom 210,000 are in Telecommunications and 175,000 are in the Postal Service. The active introduction of microelectronic technology and the adoption of new techniques like electronic mail could have a dramatic impact on the level of employment. In America, the adoption of SPC exchanges has reduced the maintenance requirements by half, and the introduction of more effective local-distribution techniques could have even greater consequences. The replacement of the postman by electronic delivery systems would displace large numbers of personnel, and have the added problem of moving the balance of labour from Posts to Telecommunications and, perhaps more significantly, changing the balance between one union — the Union of Post Office Workers, which has 211,000 members — and a union which is currently smaller — the Post Office Engineering Union, with 126,000 members.

It is obvious that any changes in this area will require sensitive and far-sighted management. Of principal concern must be the risk that the introduction of new technology will be hindered by considerations of staffing levels within the BPO. So far, the performance of the BPO in introducing new technology has not been good. The first mechanised sorting office was opened in 1966 and it is now planned that the system will be complete by 1982. This included a three-year delay between 1972 and 1975 when cooperation was withdrawn by the Union of Post Office Workers. While there must be every sympathy with the viewpoint of the worker who sees that his job may be threatened, it has to be accepted that increases in productivity, which in the long run can benefit everyone, will lead to changes in the pattern of employment, and that resistance to such changes will, at best, reduce or delay the long-term benefits and, at worst, cause a deterioration in the standard of living if other countries adopt new technology more wholeheartedly. Within this problem, the BPO must occupy a central position, since it, more than any other industry, will face a run down in staffing levels as it adopts new technology, while this technology will primarily benefit the community as a whole, and will be essential to the continuing efficiency of U.K. industry.

## Conclusions

*The provision of effective telecommunication services is essential to the development and use of information technology in the U.K. As such the BPO has a central role to play, and its management and policies need to be seen as an integral part of computing and information technology, rather than operating in isolation as at present.*

*The changing technology and role of the BPO means that its powers and functions need to be considered against future requirements. Of particular concern must be: the BPO's responsibility to develop new services like data communications which may not meet criteria for short-run profitability; the extent of the BPO monopoly in relation to electronic information; and the need for stability in services and pricing in order that the user can plan and justify long-term investments dependent on services to be provided by the BPO.*

*From the point of view of data-communication services, the most important development will be in the area of local-distribution*

175

*networks and it is highly desirable that these should be upgraded with potential data-communication requirements in mind.*

*The Post Office Research Organisation represents the largest single activity in the R & D area for information technology and its objective needs to be integrated with other R & D work. This problem is of direct relevance to the CSERB.*

*The BPO will be impacted more than any other large organisation by the use of information technology. Not only will the use of information technology change the nature of the services that the BPO needs to provide, it will also change the internal operation of the BPO, leading in the long term to a very substantial change in the level of staffing.*

## ARTIFICAL INTELLIGENCE

Because of the pathetic fallacy, artifical intelligence attracts emotional, rather than critical, assessment. It would appear that the use of computers for certain limited objectives in this area could become important in the long term.

When discussing artifical intelligence, care must be taken not to impose too anthropomorphic an interpretation on the objectives. Within the medium to long term, the results are likely to fall so far short of human intelligence that there will be a qualitative difference between specific computer capabilities, even if both are based on the same underlying mechanisms.

The term 'artificial intelligence' covers a diffuse range of activities. A common underlying theme is the attempt to create semantic models for particular subjects. Another underlying theme is the attempt to enable the computer to make inferences based on specific semantic models.

### Semantic models

It has become apparent that human understanding uses complex conceptual models for parts of the external reality. In order to provide computer-based analogues for various human activities, it will be necessary to provide equivalent semantic models for the computer. That this is the case, even for a relatively primitive activity like speech recognition, can be readily demonstrated by the difficulty the hearer has in recognising words after the thread of conversation has been lost. Because of this need for semantic models, there is unlikely to be rapid development in the general area of artificial intelligence.

The most promising short-term route to practical results is to concentrate on specific subjects, where it may be possible to create limited models, possibly of a quite different sort from those used by the human. Optical character-recognition techniques for typewritten characters are an example of this. In this case it is possible to provide an adequate level of recognition, without the need for the computer to have knowledge about which character sequences form various words, or which word sequences are meaningful. Another excellent example of this specialisation is the Dendral program, where the computer can out-class the chemist at identifying the chemical composition of an organic sample from its mass spectrogram. Within special subject domains like this, the computer may appear to operate with a high degree of intelligence, even though the more general problems of artificial intelligence have not been solved.

In the medium to long term, the most important and tractable problem would appear to be speech recognition. The recognition of a limited vocabulary with stylised pronunciation is already possible. Such systems are unlikely to have great economic importance, since they require operator training and are less efficient than keyboard input. It should not be expected that these systems could be extended to provide more general voice input, because the recognition techniques used are too primitive. The area of speech recognition is, however, promising. It is a one-dimensional process, and the entities to be recognised are well defined. To be successful it would appear that the computer will need access to a dictionary and an understanding of grammar, but that it need not 'understand' the meaning of the sentences it recognises. Since speech is the primary output channel in the human, speech recognition is a crucial element in man-machine interaction, and the availability of an adequate system would have far-reaching consequences.

The immediate prospects for general robotic systems appear poor, for the reasons already given. Again, the most promising approach would be to concentrate on limited subject domains and special solutions. There is a strong case for adding microcomputers to various forms of production equipment in order to provide a limited degree of intelligence — for example, to improve the tolerance of machines when handling out-of-specification components. Much of the current work on robotics appears over ambitious, but useful economic results could perhaps be achieved by a positive programme with restricted objectives in this area.

### Inference
Another area with long-term promise is that of general deductive reasoning. In this case, the underlying model — mathematics — is well defined, so attention can be concentrated on the inferential process of deriving theories. It would seem reasonable to expect that the computer could play a positive part in creative mathematics — the recent solution of the four-colour problem indicates the potential.

The provision of more general semantic models and general learning and inductive capability is a very long-term prospect. Judged by the complexity of the human brain, the technology will be available to support such systems long before there is adequate understanding of their design. It is possible that the development of adequate theories may have been constrained by the cost of processing power and that developments in semiconductor technology will provide a stimulus for new ideas. When results do come in this area, they are likely to lead to a rapid breakthrough, because the systems would have inherent self-improvement capability. The implications of such a development will be greater than the implications of any such other aspect of this study and should not be dismissed as science fiction fantasy.

### Conclusions
*The application of artificial intelligence techniques to well-defined and limited problem domains will have economic importance in the long term. The most promising areas are:*
> — *the use of microcomputers in sub-robotic systems to provide some degree of intelligence in production equipment*
> — *the development of general speech-recognition systems.*
> *There is also a good case for speculative research in artificial intelligence, because of its potential in the very long term, and the 'breakthrough' nature of the subject.*

# CHAPTER SEVEN

# THE IMPACT OF COMPUTING

This chapter reviews the anticipated impact of new information technology.

Computing is already far more influential than many people realise; but as the use of computing devices spreads, with the help of microelectronics, it will meet increasing resistance and its social and economic repercussions will affect the rates of adoption and adaptation.

Several case studies have been carried out, with a view to understanding these processes, and the implications of these are brought to bear upon the otherwise purely techno-economic arguments of adoption and adaptation.

Social and wider economic issues are less predictable than technical issues, and this chapter is correspondingly less firm in its projections. Nevertheless, the implication of some of these projections, particularly in relation to displacement of employment and the differential rates of adoption and adaptation, is that extremely serious issues are involved.

## THE HISTORICAL IMPORTANCE OF COMPUTING
Computing, as we know it, is less than thirty years old. Major

social, business and governmental functions are now totally dependent on it, and reversion to previous (often manual) procedures, or the dropping of functions only possible with computers, appears unthinkable. As an economic activity, computing takes place on a massive scale (see above, pp. 107 ff.), yet its sole *mission* is to provide computing capability for economic exploitation to end users. Although that capability is expressed largely in the apparently mundane form of data processing applied to commercial requirements, major advances have been made at the frontiers of science utilising computing power. All the advanced strategic activities of the major nation states — from defence communications to advanced weapon systems, from nuclear energy to economic planning — depend upon computing. So also do very many of their routine activities, from utility billing to student admission. The delivery of this capability has depended upon the use of leading-edge engineering technologies from all disciplines, incorporating scientific advances which themselves have been brought about by the application of computing power.

Yet to the man-in-the street, although he is increasingly aware of computing as computerised bills, statements or mailings arrive, the scale of today's computing is hidden. Computing is largely anonymous to him, somewhat mysterious. It would probably surprise him to learn that, in terms of absolute profitability, net assets employed and concentration of corporate control, IBM is the biggest corporation in the world. If we further declare, as we have done on pp. 31 ff., that the advent of information processing is probably more revolutionary for society even than the advent of portable motive power through the electric motor, he will be confounded — and yet, as we argue below (pp. 189 ff.), the computer age has only just begun.

On the face of it, the individual quantitative changes in computing power seem comparatively trivial. After all, any simple calculation requiring more than mental arithmetic can be mapped on to the five basic functions of input, storage, logic, arithmetic and output and the early computers did just such single-stream computation, albeit at significantly greater speed. Technological advance has increased the speed, storage/memory capacity by several orders, as well as providing the more important capacity to handle multiple-data streams, to spread the five ISLAO functions geographically and interconnect by communications, and to handle both centrally initiated computation and the more randomly initiated real-time computation. These are still quantitative changes — but when taken together and deployed imaginatively they add up to a qualitative change in our capacity to engineer activities and to respond to the activities of others.

The influence, so far, of computing on the military balance of power, and on the explanation of space and the lunar missions is clear. So also is its importance in the extension of scientific frontiers — in the unravelling of DNA and genetic coding, in the harnessing of fission, in plant and animal breeding. But we argue that its historical importance in the economic structuring of societies is currently just as great.

Those corporations and organisations most successful at the deployment of computing have prospered differentially. Since much of today's computing has continued to require complex and large computing effort, this increased differentiations has favoured the larger organisations. Since those organisations more familiar with the technology (the hardware development usually predicating software development) can better absorb and deploy computing, further differentiation in their favour has occurred. As a consequence, the efficient have been able to become more efficient, the powerful more powerful, the super-states have increased the gap towards lesser states, and the transnationals more easily transcend national boundaries. Computing has increased the pervasiveness of government, particularly in its regulatory activities, and yet perhaps has also helped provide comparative world peace by helping the creation of deterrent force in multiple depth.

The particular industrial sector where computing as this inbuilt multiplier is most clear is of course the electronic-components and capital-goods sectors, where developments are mutually stimulating and self-feeding. In association with a large defence and particularly aerospace industry (with stress on reliability and miniaturisation) the multipliers which have led to the dominance of the US in computing are clear.

Yet despite this considerable historical importance, there is no sign so far of two effects which might have been expected. Firstly, the associated service sector is still quite small and as

181

yet relatively unstructured; secondly, on the whole there has been no major displacement of labour to date. For reasons argued on pp. 113 ff. and pp. 195 ff., we expect the advent of microelectronics to accelerate the structuring of the service industry, and produce extensive potential displacement of labour.

## CASE HISTORIES

Our choice of case histories, below, was constrained not only by resources and the absence of microelectronic impact upon mainstream industries to date, but also by the need to choose a sample of activities which was sufficiently broad for balanced generalisation to be possible.

Of the three obvious microelectronics-based products — calculators, watches, TV games — watches were chosen because they could illustrate impact on trade, employment and distributive structures. The computerisation of newspapers was chosen because it was presumed to illustrate the pattern of organised-labour resistance to technical change.

Cars and textile machinery were chosen because they are major industrial sectors not only in the U.K. but in international trade, and because there was already considerable contact with the major firms concerned. Materials handling is an activity common to many sectors, of which the forklift-truck industry happens to be familiar.

It is not expected that the phenomena which can be discerned in these case studies will repeat themselves in identical ways in other sectors or activities. Each sector has its own dynamics of investment, product cycles, interdependence with other sectors, its own pattern of industrial relations, and above all its own pattern of competitiveness and structure, both in the U.K. and abroad. Nevertheless, we would expect that the findings from these case studies will be echoed in some form in all sectors. For this reason, only the major findings in each case study are given, and detail peculiar to each study is avoided.

### Newspapers and computerisation

The pattern of computerisation being adopted here was by no means the latest technology, but it represented a logical evolution

182

of computerisation advancing piecemeal over a wide range of activities, though still ten years behind potential. The displaced technology was considerably over-manned. The new technology required less manpower and at the same time offered working conditions which were cleaner, quieter and more simple.

Although the headquarters of the various unions involved had, in negotiation with the employers, agreed generous redundancy terms, the unions involved, at a local level, rejected those proposals. On inspection it became clear that the proposed national solution did not tackle the issue of the redistribution of jobs under the new technology. The basic problem is that the new pattern of skills cannot be mapped on to the previous pattern of skills, and that the previous pattern of skills had clearly-defined internal boundaries associated with different unions. Additionally, regarded as a flow process, the evolution of unions had been such that their roles appeared at multiple points in the production flow.

The union members concerned were quite informed as to the long-term potential threats to the newspaper industry, and were quite aware that full acceptance of the new technology would diminish their leverage since the new process was less multiple-step, and less time consuming. Additional anxiety was expressed about the disappearance of craft skills, and control of entry into the industry.

Barriers in this case were solely inter- and intra-union anxiety about the reallocation of jobs to workers in the absence of clear-cut criteria. The high wages demands are not high in relation to the system economics — a consequent barrier may well be the public attitude that the print unions are bloody-minded, and the resultant stiffening of attitudes.

## Cars

Early introduction of microelectronics into motor vehicles followed US legislation on exhaust emissions and threatened legislation on fuel economy. The U.K. industry interviews showed that manufacturers, component suppliers and consultants were aware of the technical potential of microelectronics but tended to believe that actual incorporation would be much slower than predicted (or had occurred in 1977).

In the U.K., the component suppliers are relatively independent of the major assembly — which is not the case in the US and Japan. In the Federal Republic of Germany, Bosch is reputed to work more closely with the major manufacturers than Lucas does in this country. In the U.K., each side stated it was waiting for the other.

Although it was recognised that sensor and actuator development lagged behind electronic capability, there was some evidence that U.K. firms were not fully aware of latest developments in these fields.

Barriers seen by the U.K. industry included a lack of semiconductor knowledge in the components industry, the familiar succession of short-term problems in the assembly industry, the problems of integrating electronic technicians into skill grades both for assembly line and product incorporation, and the absence of incentive to risk retooling for the native manufacturers, as well as the problems foreseen (unrealistically) for after-sales maintenance. Nevertheless there was a discernible interest in long-term engine development using microelectronic controls.

## Watches

Significant issues raised by this study include the prevalence inside the threatened mechanical-watch industry of Canute-like attitudes long after the tide of quartz watches were sweeping in. To a large extent the higher-quality mechanical watches, offering high margins all along the chain, were marketed almost reluctantly (using the mystique of the jeweller's advice to clinch sales). Even though the mass marketing through discount stores and supermarkets took place in cheap mechanicals first, the mainstream industry ignored the rapid turnover of the electronic hi-fi type outlets as a multiplier for quartz-watch introduction.

Even today the main watch supplier, now with digital ranges of their own, tend to ignore the potential technical development of quartz watches and point to the early, badly designed, quartz 'cheapies' as evidence of quartz failure.

At the national level the impact on employment and international trade has been large. For example, some 20-30% of the assembly labour in Switzerland was displaced, resulting in Federal intervention for the first time; and trade estimates suggest a swing in balance of payments of at least 250M$ per annum.

Barriers to this introduction were few; the market is not such that some of the early bad design would sour further sales. Indeed it would be truer to say that the barriers were negative, and that the cosy, self-satisfied, mechanical-watch industry was ripe for take over.

### Textile machinery

The incorporation of electronic controls replacing electrical controls on modern looms and knitting machines, as well as on tufting machines, has proceeded comparatively smoothly. The industry operates in a highly competitive world market, and considerable rationalisation and concentration has taken place in recent years.

Most textile-machine manufacturers employ electronics departments in their development sections and it is worth noting that whilst all volunteered that control developments had been amongst the most important in the last decade, few found it necessary to draw attention to the need for more skills in electronics.

As far as the impct on users was concerned, it was conceded that the further increase in labour productivity to be expected was small compared with that already achieved in modern factories.

In summary, it would appear that the need to concentrate on control, in all its aspects, has led and will lead U.K. textile-machinery manufacturers, in their highly competitive world market, to adopt and adapt to microelectronics in a relatively rapid and relaxed manner.

### Materials handling

As far as fork-lift trucks are concerned, the incorporation of electronic controls, particularly into electric-powered trucks, has proceeded smoothly and electronic controls for fork posture are accepted.

However, the U.K. fork-lift-truck manufacturers are becoming acutely aware that some of their foreign competitors (e.g., Jungheinrich) operate with a potential capability of using trucks with a higher degree of automation in warehouse systems. This

185

would involve a degree of computerisation and automatic truck control which would reveal structural weakness in the U.K. materials-handling sector. The lack of industrial expansion was cited as a major barrier.

## BARRIERS TO THE USE OF COMPUTING

The case studies examined, and the experience gained in discussions with many individuals from many organisations, suggest that the barriers towards both innovation and the diffusion of uses of computing and its device embodiments vary enormously between individuals and organisations.

Perhaps the most important barrier concerns awareness of the potential (and achieved) capability of computing. There is widespread awareness of particular possibilities — but amongst particular people, or particular organisations. A collective awareness of possible overall developments, which could undoubtedly impinge upon those same particular people or organisations, is exceptionally rare. And yet it is a thesis of this study that what is in the offing is a pervasive and restructuring change towards an information society. Coupled to this lack of a sufficiently comprehensive awareness is a lack of experience and a lack of technical knowledge. Incredibly, several examples have been found where individuals show good knowledge of a particular development in a field currently unrelated to their interest, and yet show total surprise when informed of the analogues and appropriate application already in use in their own field. A striking example is the manager of a factory producing a simple electromechanical product who admits to having been surprised by the sudden victory of the electronics-based substitute — although he had owned one personally for several years and found it satisfactory. It is probable that the step-function advance in electronic capability is too great for many people to assimilate unconsciously.

Another major barrier, undoubtedly, in the U.K, today especially (Dec. 1977) — is the succession of short-term problems diverting attention from technological options. Indeed, the nature of these short-term problems, which often involve manning patterns and both inter- and intra-union disputes, is such that

186

technological change (which often inevitably opens up such issues again) is basically unwelcome from this viewpoint. It can be argued, however, that this particular barrier is a consequence of another barrier rooted in current business capability. We refer to the lack of costing principles appropriate to an assessment of benefits due to technical change; normal accounting procedures are particularly inappropriate. The size of the 'cake' is unavailable, or unassessable, and discussions about the manning pattern and allocation of work are therefore carried out on a basis of pressure groups rather than on rational grounds. It is of course true, as in the newspaper-computerisation study, that technical change often results in complete remapping of skills; but, given modern union and labour leverage, it would seem that not to place this as *the* central issue is to ensure protracted dispute, as in Fleet Street.

A further barrier, in the U.K. is of course the apparently low cost of manpower and, because of the relatively depressed state of the economy, a lack of interest based on lack of demand. Since many information-based products and process machinery will not exactly correspond to such current machines, it is likely that the opportunities afforded by the expanding, buoyant markets for such products in some other countries will not make sufficient impact upon U.K. consciousness until home markets are hit by imports.

Other barriers, such as mobility of people and ideas, have been put forward; and, certainly, cultural attitudes toward change, materialism and technology appear to be of some importance. Regardless of the exact balance or profile of these barriers, it cannot be overstressed that in the case of innovation and diffusion of computing in particular, all these barriers appear to put the U.K. currently at a differential disadvantage to some of our major competitors, such as Japan, the US and the FRG. There is little doubt that the U.K. supply industry has the technical ability to formulate and design new solutions, but, since we have argued elsewhere that the health of the user industry is the dominant issue, it must follow that diffusion of innovation, from whatever source, is more critical, and here the differential disadvantages to the U.K. appear at their strongest. At this level, one great barrier to the spread of computing would appear, paradoxically, to be the apparent strength of U.K. indigenous supply industry, since the choice for many other countries, where users are concerned, is simply to buy American or Japanese equipment. Such a barrier could conceivably be turned into a positive multiplier if the U.K. supply industry applied its undoubted advisory skills to advocating best solutions regardless of source.

Another barrier commonly put forward is the availability of high-calibre personnel with a sufficiently well-rooted, wide-based appreciation of computing. The many debates on the appropriate form of 'computer science' in universities testify that the interface between tertiary education and industry, whether computer-based supply or demand, is inefficient and reflects waste, frustration and considerable opportunity costs. Summarising that debate, it appears that no clear lead has been given in the distinctions between computing as a tool, computing as currently practised and evinced in commercial data processing, digital engineering as a design process increasingly using micro-electronics, and the professional fields of numerical and systems analysis. At a time when the flexibility of microelectronics will ensure that different standards, protocols, interfaces, and other machine-based differences are of less significance in that simple solutions can be found, it may well be that considerable restructuring of tertiary (and secondary) formal education may not only be appropriate but also welcomed. Should this occur, one of the most important principles must be to ensure that microelectronics is perceived as offering not only simple effective solutions to difficult problems but also, and more importantly, simple solutions to simple problems. Economically speaking, the banal product (e.g., the TV game) often outweighs the more sophisticated product: the dilemma may well be that the banal is not often considered worthy of notice in the academic world. Other examples might well be original U.K. attitudes towards Basic, Dynamo or even Fortran as opposed to Algol.


**Conclusions**
*The main barrier to U.K. exploitation of computing technology is likely to be LACK OF AWARENESS, in a sufficiently collective of focussed way, of its potential.*

*Other barriers in the U.K. are rooted in our present industrial uncertainty and particularly inability TO ASSESS AND AGREE THE PATTERN OF BENEFITS AND ITS DISTRIBUTION.*
*The consequence for the U.K. of such BARRIERS being DIFFERENTIALLY GREATER than IN SOME MAJOR COMPETITORS is likely to be an awareness TOO LATE.*

## IMPACT BY INDUSTRIAL SECTOR

Sectors of the economy were originally delineated on the basis of their differing main technologies. Thus agriculture, industry and infrastructure gave way to mining, cement, construction, chemicals, drugs, mechanical machinery, electrical apparatus, transport equipment and so on. Unfortunately, the advance of modern technical technology has meant that different economic sectors, as delineated in most official statistics, are increasingly interdependent in that they use different mixes from a wide range of available technologies. Thus the official input-output matrix of the U.K. economy, although disaggregated to a high degree, is still based on a sector classification which with time has increasingly obscured the common and pervasive technologies, such as electromechanical assembly for light equipment, or office technology. For example, metal shaping is common to many industries, as is transformation of energy from one medium to another before final use, as is manipulation and distribution of information.

Nevertheless, by looking at the characteristics of processes or products where the need for some form of computing is clear, it is possible to start gauging the probable impact of the information revolution upon different sectors.

Such process characteristics are:
— presence of monitoring, surveillance, leading to diagnosis by comparison, leading to predetermined or computed response, and where these requirements are needed frequently or continuously, a responsiveness needed quicker than human, environment alien or hostile to human, need for responsiveness free from human interference whether malicious, foolish or misguided, need to send local report elsewhere in quick time, where

189

actuators can be manipulated electrically, where these requirements are repetitive or sequentially repetitive. Such product characteristics are:

— mass production, incorporating electromechanical controls, already incorporating logic, using light current power, and where the cheapness, reliability, and miniaturisation with reduction in space and power, and where ruggedisation can be deployed. Or small batch currently incorporating specialised controls displaceable by digital logic, preferably programmable to a microprocessor or microcomputer. Products capable of attachment to existing products such as TV, radio, hi-fi, telephone, computer peripherals.

Industrial characteristics (e.g., at the firm level, or at the establishment level) favourable to rapid innovation and diffusion are:

— familiarity with light electrical controls or electronic controls already, exposure and familiarity with product offerings in more dynamic markets, and those exposed to such competition. Possibly those with a good track record of innovation or adoption.

Such a list of characteristics, both product and process, cuts well across the normal sectoral classifications and even within a firm it may well be that opportunities for change are not fully visible from any one vantage point.

So far we have considered very largely the manufacturing side, although, as pointed out below (p. 195), the labour impact is also very considerable on the design, installation, maintenance and repair side. But even in manufacturing sectors the directly associated service activities of procurement, stock control, and clerical work associated with orders are perforce greatly simplified when production lines are 'simplified' or automated — the high ratio of 'clerical' work to direct 'production' work even in manufacturing firms is greatly under-appreciated. For these reasons we expect the impact, not only in labour required but in managerial skills, to be greatest in the 'overhead' structure of firms — in other words, in the 'white-collar' area.

Another — and in many ways potentially the most useful — way of looking at impact is to replace the sector concept by an

190

occupational concept. To illustrate this approach, the US Bureau of Labor Statistics 1970 data on occupations, and the U.K. 1974 Earnings Survey data on occupations have been analysed.

| Information handlers | 165 occupations totalling | 29.4M |
|---|---|---|
| Industrial workers | 108 occupations totalling | 15.1M |
| Service industries | 120 occupations totalling | 26.3M |
| Agricultural workers | 8 occupations totalling | 3.2M |
| | | 74.0M |

Grading each individual occupation in turn as high, medium, low or zero risk is 50%, 25%, 10%, 0% potential job loss over the next fifteen years:

| Information handlers | 7.8M | = | 26.5% |
|---|---|---|---|
| Industrial workers | 3.3M | = | 21.8% |
| Service industries | 2.3M | = | 8.7% |
| Agricultural workers | 0.1M | = | 3.1% |
| | 13.5M | = | 18.2% |

Reducing the potential job-loss rates to 25%, 10%, 1%, 0% for high-, medium-, low- and zero-risk categories respectively, a projection of some 7.2% weighted overall loss is obtained.

On a similar basis, using the higher rates of possible job-loss, the U.K. data (based on a total of 240 occupations, data drawn from the New Earnings Survey 1976, Part D) projects a displacement of some 16%. Obviously, such projections are based on a view as to which occupations are at risk. The major classes are proof-readers, library assistants, mail carriers, telegraph operators, draftsmen, programmers, accountants, financial advisory, administrators, secretaries, billing clerks, keypunchers, cashiers, filing clerks, meter readers, shipping clerks, TV repairmen, plateprinters, telephone repairmen, light electricians, machinists, mechanics, inspectors, assemblers, operatives, material handlers, warehousemen, sales clerks, stock clerks, compositors.

Clearly, impact on any scale like this will place great stress on many firms, and considerable industrial restructuring would be probable. This might well take the form of new firms expanding, whilst old forms or subsidiaries collapse. In many ways new firms, unsaddled with the burden of older processes, plant and personnel, can take advantage of potential increase in efficiency more easily, and even deal with a changed requirement for levels of both skilled and unskilled manpower more easily.

## ECONOMIC ROLE OF THE COMPUTING INDUSTRY

### Growing pervasiveness of computing

Probably the best analogy in previous and other technology for a similar economic role to that which will be played by the availability of pervasive computing devices is that of the fractional horse-power or small-scale electric motor. We refer here not to the modern linear motors or other recent developments but to the pervasive squirrel-cage motor.

One of the problems in discussing the economic role of computing, whether in the supply industry or in the user industry, is that the boundaries of the computing sector are becoming indistinct or blurring into other sectors. Thus many office machines, telephonic exchanges or switchgear data-collection systems, sorting machinery, incorporate computing principles — a problem well known to those attempting to interpret production or trade statistics in this field. Yet, today, no one would talk explicitly of a fractional electric-motor sector, although such motors are one of the most common technological artefacts in our society — just how common is shown by the fact that the typical AB household has on average some eighteen motors in it.

### Supply-industry size today

The bare statistics of todays' computing-supply industry in the U.K. are approximately as follows: at year end 1977, some 9500 general-purpose installations with central processing power greater than that of IBM System 3, and an installed value of some £38B. However, such data is particularly misleading in relation to the rapid changes in pure computing power available at increasingly lower-cost thresholds.

Future projections vary enormously, but we consider it likely that future growth rates will asymptote down to or even fall beneath (or go negative) those of national growth levels. This is to recognise the fact that considerably increased power will be available via LSI circuits at reduced costs, even taking into account the vast expansion of usage applications envisaged. This is not to belittle the importance of the supply industry but rather to point out that economic importance does not necessarily equate with size or turnover.

## The future of the general-purpose mainframe

We have postulated also that many computing devices will be directly incorporated into products, whether consumer goods, durable goods, or capital equipment, and be regarded as much a part of that good as the electric motor is of a vacuum cleaner. It follows logically that, with the direct provision of intelligent and local responsive power to many information-handling devices, we expect the nature of the hardware supply industry explicitly recognised as the computing industry to change, in both direction and scale, particularly as far as large mainframes are concerned. Given the momentum of implementation and the fact of major investment in large configurations incorporating at least one general-purpose mainframe today, it is likely that the demise of the large general-purpose mainframe will not be clearly visible for another seven to ten years, but we assert that it is certain, and that the change to customised sets of processors, balanced in relation to mission, is already in train. Obviously individual sub-mission-oriented processors (cf. today's minicomputers) will be a major part of hardware output; but, as we saw above (p. 99), the concept of a computer system is giving way to some form of system of access to computing power.

## Software functions in future

In relation to software, we would postulate that the distinction between systems analysis, systems programming, applications programming and coding is likely to change, with considerably more attention being paid to an understanding of the functional

193

elements, and their system relationship, of the application being studied. Since we are arguing that the trend must be towards dedicated application, it is clear that the bulk of implementation work will be towards such functional analysis, location of available computing circuits and devices or determination of optimum design — with today's stress on software and programming totally played down. That this has not yet happened is due to the fact that available microprocessors are simplified copies of large-scale processors built up through a complex evolution of clumsy technologies — the simplification unfortunately taking the form of rendering unnecessarily complex, and high, languages difficult to implement via restricted and clumsy order codes.

### The future implementation house
We anticipate that such implementation houses will displace today's software houses and, for analogous reasons to those increasingly pertaining in software today, will be based upon a range of specialisations in applications. Details of implementation techniques and choices, being less complex than today's software solutions, will be even less protectable in a security sense, and such implementation houses must of necessity go for the largest possible exploitation markets.


## ECONOMIC IMPORTANCE OF THE USE OF COMPUTING

### The economic multiplier provided by computing
All productive sectors, whether manufacturing, commercial, financial, service or governmental, will increase their productive efficiency substantially, although with potentially large displacement-of-labour consequences. On the assumption that the nation-state entity, together with redistributive mechanisms to cater for purchasing-power inequalities, remains the chief form of social organisation, the major economic importance of the use of computing must be to preserve the trading balance of the nation-state vis-á-vis its competitors in trade, and for this reason the chief economic role must be to preserve competitiveness.

Given that identification, location and assessment of competing products, and ordering, distribution, support and maintenance of products will all be enhanced by the spread of computing, it follows that competition will increasingly be based upon direct comparison of performance/cost ratios, *ceteris paribus*.

## Computing and industrial economies of scale
Given that the manufacturing sector (unlike the commercial or financial sectors) of the U.K. is absolutely smaller than that of our current major competitors and many emerging competitors (and must remain so for some time), many of the economic factors affecting cost are adverse in such a comparison. As mentioned above (pp. 118 ff.), a large multiplier to R & D productivity will become available from microelectronics; perhaps this activity, being brain intensive, offers less comparative disadvantage than others to the U.K.

During the course of this study the overwhelming importance of the microelectronics revolution to comparative economic performance has become increasingly accepted, although acceptance is still restricted.


# EFFECTS ON EMPLOYMENT

## Vectors of change — and barriers
We have postulated the availability of microelectronic devices which both change the pattern of computing as recognised today and spread computing devices into products and processes associated with manuacturing, service and everyday life. In assessing the impact of this change we are aware that even hind-sight studies of past technical change are frought with conceptual difficulties and multiple interpretations. We propose, therefore, to raise the issues that will affect employment patterns in the U.K. and elsewhere in a structured way, reflecting the vectors of change and the barriers to it that have been put forward earlier. Exactly how these vectors will interact to form the matrix of part of our future society will depend not only on the capability of U.K. industrial and governmental policy to influence those vectors, but more importantly on the influences brought to bear by the major companies and governments outside the U.K.

**Impact on assembly**

Microelectronics-based circuits can be substituted for a large proportion of assembled devices commonly used throughout the economy — such as electromechanical meters, switchgear, complex assembled circuits such as sub-systems of radios and TVs, and the other measurement, monitoring and control devices currently utilising any mix of mechanical, electromechanical, electric, discrete or integrated-circuit technology. It is a characteristic of the production technology of such current devices that the labour-density pattern of activities involved in production is roughly pyramidal in shape, with relatively large numbers of low* skills at the base activities, and smaller numbers of higher* skills near the top of the pyramid. A typical distribution could be:

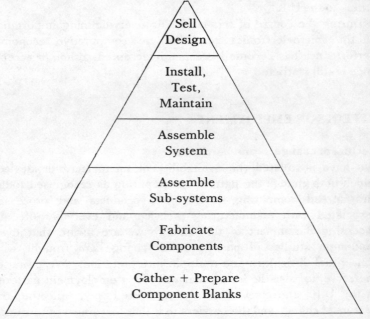

This activity density pattern is product based, not necessarily firm based. Thus for example amongst car manufacturers British Leyland is chiefly a 4-5-6 company, but with agents carrying out

* By contrasting high/low skills we imply only skills requiring more/less training, or adaptive/rote based.

196

much of 5; levels 1, 2, 3 in this case are networks of suppliers who often supply other car assemblers also. At a more elemental level, screws, rivets, paxolin boards, wires, discrete components, are bought in for electrical-circuit assembly by a 4-5-6 electrical-product manufacturer.

In order to control this pyramid of activity, managerial and technical skills are also required. A characteristic of both is that with this type of activity the span of managerial or technical skill need not extend through the whole multilayered pyramid: it is currently sufficient if such skills bridge adjacent layers:

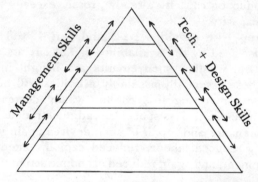

However, the nature of both the design process and the production of a microelectronic circuit is such that this structure is totally altered. For example, the next figure shows the part of the pyramid totally displaced by an LSI circuit, and the corresponding requirement for span of managerial and design skill. It also indicates diagrammatically the nature of the labour displacement effected, when such substitution is possible.

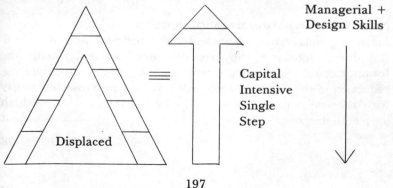

From this conceptual diagram — plus the earlier point that such devices are cheaper and more reliable than the devices they displace, plus the additional point that they are individually less material- and energy-intensive — we can start to structure the likely employment and social impact.

### Negative multiplier inside electronic firms
The semiconductor suppliers themselves, will, for equivalent output, become less labour intensive, and, at full production with a stable production technology, become less capital intensive. Present production capacity already grossly exceeds foreseeable absorption capacity.

Those firms with activities 1-2-3 and part 4, will have such activities displaced by the availability of LSI circuits embodying such activities, whether such circuits are bought in or manufactured. These essentially assembly activities will be displaced by an activity involving less capital, vastly decreased labour input, less supervisors, less space, less administration.

The maintenance and repair of such devices again will require substantially reduced labour, reduced capital content for diagnostic equipment, and again reduced economic activity.

### Negative multiplier for suppliers to electronic firms
Since the products and process equipment incorporating microelectronic elements are more powerful in their performance than their predecessors it follows that, for the same performance or output requirement, less product or equipment is required, and that there must be a corresponding drop in the labour required to replenish the stock of products or equipment in use.

### Negative impact upstream to such suppliers
Since the substituting circuits are less material intensive, and less energy intensive, they are also less handling, machining, forming and fabrication intensive, and there is a corresponding reduction in the need for equipment to perform those functions, with the concomitant reduction in labour required to produce such equipment.

198

### Negative impact on clerical and administration sector

The widespread adoption of microelectronics-based products in the service sector (e.g., office automation or today's 'computing') may be expected to have large labour-displacing effects also. Routine copying (whether straight copying or copy typing) and filing are both labour-intensive activities. The trend towards local processing in computing, and the direct acquisition of associated data through databases, will hit the cleric-based operations currently associated with today's patterns of organisation and correspondingly reduce the middle- or low-level managerial skills used to initiate or monitor such activities.

### Relative absence of new products

It is hard to envisage new products, whether consumer or capital goods, which are of sufficient volume and value to absorb the labour displacement put forward, at least in the same timescale. Studies of the labour-displacement effect of electronic watches versus mechanical watches, even of the apparently new product of the portable calculator versus the reduced demand for electro-mechanical desk machines, suggest that the negative labour effect is much higher than the relatively trivial positive labour requirement for, say, electronic TV games. Labour might however be absorbed in the provision of quite different goods and services after an adjustment process.

### Competitiveness of national economies and differential adaptation

The analysis so far has approached the issues as though the impact was on one economy. However, since the world markets (largely, advanced countries) consist of competing economies, or rather competing firms with differing national bases, any disparities between countries in their rate of adoption and adaptation to the microelectronic revolution are likely to result in differential economic multipliers between countries. Thus a country that successfully adapts and adopts may *ceteris paribus* have to absorb the consequent structural change first and prosper (economically) and therefore as regards employment, second, through substantially increased competitivity in its exported products, services and commodities.

### Likely pecking order of national economies, ranked by rate of adaptation

Factors likely to effect the differential rate of adoption, adaptation and absorption of microelectronics-based products and services can be listed as follows: comparatively high economic activity, local presence of strong microelectronic capacity, local presence of strong organisations downstream which can use microelectronics, strong marketing and distributive networks in export markets. From these viewpoints, it would seem likely that Japan, the US, the FRG and the Netherlands (albeit for different reasons) will lead the way, with Canada, Sweden and France following in advance of Italy and the U.K. It is perhaps ironic that the U.K. with its highly organised distributive systems, particularly in consumer goods (especially durables), and its highly organised service activities, may well receive the relevant microelectronics-based products next in order to the innovating country.

### Types of labour affected

In summary, the types of labour to be affected by such displacement can be characterised in various groups: assembly workers (chiefly rote-skill based and working with fairly complex but not very high-value products), repair/maintenance workers, low-skilled (or even high-skilled but rote-based) clerical workers. Obviously, lack of competitiveness in a U.K. firm compared with an overseas competitor in an open market would affect all categories of labour.

### Skill structure of personnel responsible for adaptation

To support the adoption and adaptation of microelectronics, considerable changes are necessary to the skill structure of the adopters and adapters. We believe that it is arguable that the level of software skills required will be less than that associated with today's unnecessarily complex computers. Less arguably, we consider that implementation and adaptation skills will require an approach to products and processes which aims at a functional analysis of functional patterns. Such an approach to determining functional 'what' steps across many industries and products is only common amongst consultancy and service industries and is not currently reflected in vocational training, which is somewhat industry oriented.

## SOCIAL CONSEQUENCES

### General levels of unemployment
Should the labour displacement consequences put forward above (pp. 195 ff.) take place in a society experiencing low growth and balance-of-payments problems, we may be contemplating levels of unemployment c.10-15% of the workforce and more. Such a level is roughly that of the trough of the 1930s depression, but with the difference that today's society is more highly organised.

### Differential social impact of labour displacement
The acceptability of such a level of unemployment is not clear, particularly since it can be expected that white-collar workers as well as blue-collar workers will be affected. The displacement of most of the female labour force (often 'twilight' shift workers) whose income has often become an essential part of the family income, may also cause unpredictable reactions for feminist groups. Yet since any labour impact is likely to occur at factory or establishment level, and likely to discriminate against particular groups of workers, it is not likely that spontaneous demand for work sharing and rearrangement of work-earned income will emerge. Attempts to 'average-out' work under redundancy pressure have not been noticeably successful in recent years.

### Feedback to political issues
The extent to which redistribution of purchasing power, at considerably higher unemployment levels, is socially acceptable is debatable. Clearly, there is massive scope for increasing the efficiency of transfer payments through efficient computerisation, although that too might swell the unemployment total. It must be admitted that there is no basis, in experience or by analogy, for hypothesising the social consequences of unemployment rates of 10% or more in an advanced society of the U.K. type in the context of the 1980s. What seems likely is a demand for job protection and job security, a political acceptance of those needs, and a creeping protectionism, with its consequent economic stagnation.

201

## Cumulative, as opposed to sudden, impact

The nature of diffusion and innovation using microelectronics is likely to be via the penetration of and dog-fighting over individual markets, one by one but sequentially. Paradoxically this implies that the scale of the problems caused is always at that time relatively small, localised and 'explainable' in terms of 'special' factors. The accumulative effect of the microelectronic revolution may well not appear until too late for effective industrial-strategy planning.

## Trend towards regulated societies

The implications of labour displacement, in our view, dwarf the concern about the tendency towards '1984-type' societies, and the issues of privacy. Indeed it may well be that, for socially acceptable forms of equitable distribution of purchasing power (however 'equity' is politically and socially defined), there may be a need to introduce more comprehensive record-keeping of individual/family incomes, commitments and needs.

## Political instabilities at national level

Increasing disparities between standards of living in the advanced OECD countries is likely to trigger political instabilities in both the military and the trading relations of the Western countries. There seems little doubt that continued recession (at 1977 levels) may result in the spreading of Eurocommunism beyond France and Italy, and many West Germans would hesitate to predict the political future of the FRG if further significant increase in unemployment occurred there. The only reasonable assertion that can be made about advanced societies meeting stagnation is that political activity seems to take the form of moving away from traditional alignments, with splintering of interest towards local or sectional interests, an apathy towards wider affairs, and general fracturing of social identity.

## Japan as a special case

The one OECD country which may be expected to adopt and adapt, has reasonably high expectations of continued growth and, for cultural reasons, will accept such technical change is Japan. The peculiarly Western concept of man versus machine does not exist in that particular psychological form in Japan, as is evidenced by the total lack of debate on the privacy or '1984' issues. The previous three 'discontinuities' of Western man put forward by Freud — viz. the Ptolemaic, Darwinian and Freudian revolutions — likewise find no parallel in Japan.

## USE OF COMPUTING IN GOVERNMENT

At all levels or in all structures of government whether central or local, concerned with national, regional, local, individual or aggregated individual, the impact of the forthcoming revolution in information handling and manipulation will affect both the nature and capability of government, and the efficiency of the government processes.

For example, at the statistical level, whether concerned with census of population, output, earnings, trade, taxation, or other indicators, the process of data gathering whether through form filling or by product data capture can be made more appropriate, involving less clerical effort, and cross related without both massive effort and its associated delays. The consequences of the decreased delay alone is significant at the centralised level, where too often aspects of the economy are not available to the same time base. Thus trade data by sector can be related to the lags and leads in currency commitments, and seen in the perspective of a current input/output matrix of the economy. Thus the capability of following and monitoring the dynamics of economic perfor- mance can enhance the ability to understand, both at macro and at micro levels, and thus to exercise control on both coarse and fine tuning of the economy. Since it is now accepted that, in the recent years of both stop/go intervention and the discontinuities of economic shocks exemplified both by the oil crisis and the many currency upheavals, the inability to assess current patterns of events is a major and costly handicap, it would seem that in this field alone major economies are possible.

This increasing ability to co-relate and cross-access data will undoubtedly raise issues of confidentiality at the business level, and privacy at the individual level. Provided that full use is made of the relatively cheap capability to control access to data, to register such access, to regulate access, it can be argued that modern interdependence both at the business and social level is such that a high degree of data availability is desirable and acceptable. For example, consideration of the consequences of failure to carry out cross access between different organisation files, as in the Maria Colwell case, would outweigh many considerations concerned with too much access to personal data. At the business level, the enhanced ability to assess or seek new commercial markets or opportunities for relationships in a more efficient way would certainly be attractive to the more enterprising firms and organisations.

Improvement both of data availability and its presentation will result in much governmental clerical work, highly patterned and large scale, being ripe for the process of office automation discussed to some extent elsewhere in this report. Much low-level governmental clerical work (e.g., of the DHSS type) could be transformed from being dull, repetitive and soulless into interesting, specifically helpful and thus professionalisable work. The prospect of reduced demand for clerical labour must in this case be weighed against the potential of using the 'redundant' force to follow up and extend more effectively the supportive and eventually productive caring functions. Considerable changes are required anyway in attitudes towards work as the, currently, only available definition of social utility, and improved ability to help relocate talents in a socially productive way would probably reduce demands on the overloaded and creaking supportive services with massive net gain. Should this argument not be immediately convincing, we would advocate that government, by its nature, is in a unique position to carry out comparative experiments with the ability to take identical sets of offices and negotiate planned experiments on differing degrees of office automation, manning patterns, and use of spare manning capacity.

At the national level also, tremendous advances can be expected in the mechanisms of formulating legislation, improving its consistency and universality, and in its updating. The presentation of, and awareness of, rules and regulations is apparently a cause for concern, and yet the presentation of structured branched description of rules and legislation would appear particularly appropriate and simple for microelectronic based presentation.

For example, Viewdata, whilst not necessarily the most appropriate medium, could reduce the present cumbersome 3000-page VAT manual to a simple search routine. The experiments of IBM U.K. Scientific Centre, in conjunction with local DHSS offices, in placing simple question-and-answer-based search routines on social-security availabilities via terminals in public Post Offices indicates very well the massive scope for such approaches.

Again at the national level the availability of wider and better data provides a basis for the better consideration of different fiscal systems, not least in the permanent debate between centralised and localised tax raising. Should this possibility of increased government ability to change procedures appear to threaten perpetual change, the possibility of using national referenda cheaply and effectively can be considered.

## Conclusions

*The impact of computing on government can affect both the NATURE and the EFFICIENCY of government processes.*

*Economic management can become more efficient, chiefly because the data will be more RECENT and CO-RELATABLE.*

*Governmental clerical work is particularly appropriate for OFFICE AUTOMATION and for PRODUCTIVE REDEPLOYMENT of surplus staff.*

*Legislation can be made both more EFFICIENT and ACCESSIBLE.*

*The potential will exist for more efficient forms of political PARTICIPATION.*

# CHAPTER EIGHT

# POLICY PROPOSALS

### THE ROLE OF THE COMPUTERS, SYSTEMS AND ELECTRONICS REQUIREMENTS BOARD (CSERB) IN GREAT BRITAIN

The scenario presented in Chapter 3, of a series of changes leading towards the information society, implies a very different view of computing and its role in society from that prevalent when the CSERB was created. Since the CSERB is itself concerned with influencing the future by its investment policy, it is pertinent to consider whether the role of the CSERB needs modification to take account of this change of view. The fate of the CSERB should be of interest to other countries who are also creating official bodies to shape and monitor future policy regarding microelectronics.

### Terms of reference

The CSERB is one of a number of research-requirements boards set up to implement the customer/contractor relationship recommended in the Rothschild Report of 1971 to control government-funded R & D.

The terms of reference for the CSERB, as revised in 1976, are:

To determine, subject to the agreement of the Secretary of State for Industry, the objectives, composition and balance of the Department's intra-mural and, where appropriate, extra-mural R & D programmes and the broad allocation of funds in this field to further the practical application of technology and to advise the Minister on other matters relating to science and technology within the Board's field of interest.

The domain of the Computers, Systems, Electronics Requirements Board shall include R & D requirements as well as other scientific requirements:

1. computers (hardware, software and peripherals)
2. computerised systems and the application of computers (automation, CAD, etc.)
3. telemetry, data collection, data communication
4. man-computer interaction
5. electronic components
6. electronic instruments for scientific, medical and industrial purposes (and such other matters as shall be allocated to it by the Chief Scientist's Chairman of the Boards Co-ordinating Committee under the Chairmanship of the Chief Scientist, of the Department of Industry).

Against the view of the future that has been presented, the terms of reference for the CSERB can be seen to have a major strength, a major weakness, and a serious ambiguity.

The strength of the CSERB lies in its integration of responsibility for the electronics and for computing. Electronics will become increasingly dependent on the techniques of computing as electronic components become programmable devices. This convergence means that the two subjects need to be considered from a common viewpoint.

The weakness of the CSERB is that it has no responsibility for telecommunications. The convergence between computing and telecommunications directly parallels the convergence between computing and electronics, and there is a need to see all three as the basis of a single information technology. The development of telecommunications and telecommunications services will become increasingly dependent on the techniques of computing to an extent which transcends the present CSERB responsibility for data communication. Underlying this weakness in the terms of reference is the fact that the reporting paths within government for electronics and computing are very different from those for telecommunications, which is the responsibility of an autonomous body, the BPO.

The ambiguity in the terms of reference lies in the extent to which the CSERB is concerned with the use of computing. Although the applications of computers lies within its remit within the conventional context of general-purpose programmable computers, it is not clear externally to what extent the CSERB has a responsibility to promote applications outside the conventional scope of computing, for example in the motor car. Again, this difficulty reflects the way authority is organised outside the CSERB. While the CSE division of the DoI has a responsibility to the computer industry, it has no direct responsibility towards promoting the use of computers, except insofar as this promotion benefits the computing industry itself. It can be argued that the responsibility for the exploitation of computing lies within the relevant industry, and its associated government structures. However, in the case of an emergent technology, like information technology, there is a clear professional responsibility to ensure that the technology is properly disseminated and exploited. This responsibility must, and should, cut across established departmental responsibilities.

**Computing policy**
So far, the CSERB has operated in response mode to R & D proposals generated within the GREs. Such a mode of operation is unlikely to achieve efficient utilisation of the resources available, and the CSERB needs to adopt a more positive role, directing R & D according to a future-based policy.

The CSERB is concerned with investment for the future. As such its decisions should be based on a coherent view of the future, and be part of an integrated policy. At present no such view or policy exists and, without these, it is difficult to see how the CSERB can function effectively. For example:
— A decision to carry out R & D on specific types of computer architecture as is happening at RSRE is only of value if it is expected that a relevant manufacturing industry will exist to pick up such R & D.
— Decisions on the role and function of the NCC are dependent on the extent to which promotion of the use

208

of computing is seen as a worthwhile aim in itself. Within a formal policy, there is the risk that decisions made by the CSERB will lead to the waste of public money or, at best, create a *de facto* policy by default.

It is not the responsibility of the CSERB to determine computing policy. The CSERB does, however, have a responsibility to press actively for a policy framework within which it can operate, and it should play a constructive role in the formulation of such policies.

## Conclusions

*The CSERB should consider how telecommunications R & D can be integrated with existing computing and electronics interests, and recommend appropriate changes to its terms of reference.*

*The CSERB should consider the extent to which the use of information technology needs promotion, and recommend appropriate changes to its terms of reference.*

*The CSERB should press for the formulation of a policy on computing to provide a framework within which it can deploy its resources effectively.*

## AREAS FOR INVESTMENT

The purpose of the current pattern of investment by the CSERB is somewhat unclear to the authors. Rather than examine this pattern in detail, it was decided to consider the pattern of investment best suited to the changing situation, without examination of the practical constraints which may limit the action of the CSERB.

Chapter 7 shows that, from the point of view of the U.K., it is the use of the technology which has the dominant economic importance, and that it is here that investment should be concentrated for the best effect. The proper use of the technology will affect every aspect of industry and commerce, and as such is essential to the future economic well-being of the U.K. The computer industry, by contrast, is only one sector of the economy. It is a sector which may exhibit instability, which is unlikely to grow rapidly where the U.K. is not well placed to compete, and where there will be strong competition from other major countries.

At present, the investment pattern of the CSERB is largely directed towards activities in support of the industry, rather than in support of the use of computing. It is recommended that this pattern should be reversed.

In the following sub-sections, individual areas for investment are discussed. These are brought together on pp. 217 ff., in the section on organisation of research.

### Support of R & D within GREs

The case for supporting R & D leading towards hardware-oriented products needs to be reconsidered. The relevant parts of the computer industry are large enough and have sufficient technical competence to carry out this work themselves. To carry out such work in GREs only reduces the funds that can be made available directly to industry, and subtracts from the pool of expertise available to industry. It is also questionable whether the results can be transferred effectively to industry. The CSERB should examine critically all the activities it has in this area to see if they can be justified against the alternative of direct support to industry.

The situation with respect to software-oriented products is different. The software industry is fragmented and under-capitalised. Many of the companies are small and do not have the resources to carry out necessary R & D. Although it may act as a palliative, the provision of R & D within GREs is no solution to the structural problems of the software industry. Responsibility for initiatives in this area lies outside the ambit of the CSERB, and any action by the CSERB should be subordinate to a more general policy.

### Support of R & D within the computer industry

The CSERB provides support for industry R & D through the ACTP and Software Products scheme. In each case, the authority of the CSERB is indirect, being delegated to the Advanced Computing Techniques Products (ACTP) Committee (largely operated by the NPL) and to the NCC respectively. It is also possible for CSERB to provide support directly to industry, while the DoI provides such support independently through the launching-type aid for ICL and the microelectronics scheme.

The rationale of such support can be questioned:

— R & D costs are acknowledged to be only a small part of introducing a successful product to the market. Excessive encouragement of R & D may be the cause of the well-known syndrome of lots of good ideas and few profitable products.

— The decision about areas of investment and development are made by technology-oriented personnel rather than commerce-oriented personnel. This seems to be quite wrong, and could represent another factor in the non-commercial orientation of U.K. companies.

— Control over such expenditure is extremely difficult and expensive. Because the investment does not lead to a direct return in the normal company accounting process — and indeed, there is sometimes no tangible result — monitoring is required at a detailed technical level. This is often beyond resources of the sponsoring body, because of the specialisation involved.

— Within a company, there is often no distinction between the pound note spent on research and the pound note spent on sales. It is impossible, therefore, to ensure that money provided for R & D is used for this purpose, and does not merely displace internal funds to be used for some other purpose.

There is a strong implication behind R & D investment that such activities will stimulate industry. This is an attempt to catch a mackerel with a sprat, avoiding the large-scale investment which is necessary to promote the industry effectively.

It would seem more appropriate that investment in the computer industry should be considered on a proper banking/commercial basis, and that where such investments depend on R & D, specialist technical advice should be provided on the feasibility of the proposal. Where necessary, the investment should be contingent on technical results. In such a case, there would be a residual role for the GRE's to provide a monitoring and technical assessment service, but no role for direct investment by the CSERB.

211

**User-oriented R & D**

As Chapter 7 has indicated, the primary barriers to the use of the technology are unlikely to be technological. They will arise from a lack of technical awareness and expertise, from a reluctance to invest in advanced technology and from the changes that the use of the technology might cause. For these reasons it does not seem that investment in technology is nearly as important as investment in use.

**Support of user R & D**

The support of pilot applications in various industries has been suggested as a method of promoting the use of the technology. There are practical difficulties in doing this, because of the competitive nature of the companies involved. There is also little evidence that technology transfer has been successful in this situation. Again, the U.K. can boast a considerable number of firsts in the application of the technology — for example, in materials handling — without these having generated any general benefit.

While the primary barriers to the exploitation of information technology seem to lie elsewhere, the demonstration of the feasibility of using the technology in specific circumstances would seem to be worthwhile. To be effective, however, it is not the pilot application which is important, but subsequent chinese copies for other users. This suggests that it is not the user who should be supported, but an independent system house which can apply the expertise it has gained in further applications. To be effective the support should be arranged so that the service companies involved make no profits from the first installation but can make substantial profits from subsequent installations. The simple approach would be to provide the service company with an (interest-free) loan repayable in a fixed term, and with no attempt to take royalties from future sales.

Decisions on this type of support need to be taken in the context of the relevant industries and the commercial viability of individual companies in the service sector. Neither the Software Products Scheme, nor the CSERB itself is the appropriate vehicle for such a policy. The role of the CSERB should again be to support a group to provide a monitoring and assessment service.

It is suggested that the CSERB could take three positive steps in this area. It should:

- initiate studies on technology transfer in cases where this approach has already been tried, to determine the problems and likelihood of success
- draw the attention of the DoI as a whole, and not merely the CSE division, to the possibilities in this area
- in default of any better mechanism being established, pursue support of specific industrial uses.

### Semiconductor technology

While the future stability and profitability of the semiconductor industry must be very questionable, there can be no doubt that this technology has the greatest importance militarily and will become basic to almost every aspect of industry and commerce.

The question whether the U.K. should attempt to compete in the semiconductor industry lies outside the brief of this study. It would seem, however, that there is a clear case for maintaining expertise in this technology, and in default of any other mechanism the responsibility for this must fall on the CSERB.

The arguments in favour of maintaining competence in the technology are:

- the strategic nature of the technology (in both the economic and the military sense)
- the need to understand its potential and exploit relevant developments
- the possibility that, as the technology stabilises, there will be good competitive opportunities in this market.

As well as maintaining expertise there is still great potential for semiconductor development and it is suggested that research in the following areas could lead to long term pay-off:

- large-scale displays, which will be essential to low-cost access to information systems
- sensor technology, which will be required for access to microcomputer control
- advanced semiconductor phenomena particularly using the wave characteristics of electrons.

The sum required to maintain a presence should be quite low, of the order of £5M p.a., which would seem to be a very good

213

insurance policy and speculative investment.

### Languages
The central role of languages, and the failure of natural mechanisms to promote language development in a beneficial direction was discussed in Chapter 4. This is an area where government action is necessary to promote research, development, standardisation, promulgation and exploitation. There is a good case for creating a centre of expertise which can exert its influence to ensure that developments in language benefit the U.K.; the cost of this might be of the order of £2M p.a.

### Monitoring
The total investment by the U.K. in R & D is small in comparison with countries like America or Japan. Both these countries, however, maintain an active monitoring capability on foreign activities in order to ensure that they are aware of significant developments. Within the U.K., some money might better be spent on providing an equivalent monitoring service of foreign technical developments.

### Standards
The U.K. is seen primarily as a user, rather than as a supplier of information technology. Good standards are in the interest of the user because they improve the utilisation equipment. The CSERB should, therefore, be prepared to spend money to establish better standards.

The primary standards for the future will be in the area of language, since most other standards will reduce to, or be related to, language standards. Standards will also be relevant in the area of physical interconnection and components. Apart from languages there are three areas where positive and good standardisation can be of benefit to the U.K., and where the U.K. lead might enable commercial advantages to be gained:
- standardisation for intercommunication of word-processing systems
- general communication standards
- interface and interconnection standards for microcomputers and associated circuits.

There are a number of actions which the CSERB could take in the area of standardisation:
— the study of requirements for standardisation, and types of standards necessary
— pressure to improve international standardisation procedures
— coordination of U.K. activities in standardisation, and presentation of U.K. stance in international standardisation
— support of studies and evaluation for ongoing standards
— financial support to individuals and groups involved to standardisation.

The CSERB should also support agreed standards (as is currently done with Coral). Such support should involve maintenance of the standards, promulgation and training.

## Policy and technical advice

Because of the rapidly changing nature of the technology, and its key role for the future, there should be far greater emphasis on the assessment of developments in the technology, and of its impact. The support of long-range policy studies would do much to alert government and industry to the potential for change, and to provide a useful basis for policy decisions. Almost all the areas within this current study warrant treatment in greater depth. Of particular concern must be the impact of the larger developments on industrial and social structure, and the implications for employment and education. It is thought that the consequences of the following developments merit particular study:
— future structure and role of the computer industry
— the integration of computing and telecommunications, and the implications of this on the role of the BPO
— future developments of telecommunications services
— the impact of the microcomputer on the competitive position of the U.K. products in world markets
— the effect of word processing on office organisation, efficiency and employment
— development of electronic mail
— development of electronic money
— the requirements for improved local area telecommunication distribution

— the future of electronic information services.

## Use of computers in government

In the long term, the development of electronic information will have greater implications for government than for any other part of the economy, since a vast part of government apparatus is concerned with the acquisition and dissemination of information. The positive adoption by the government of advanced technology in this area would provide a direct stimulus to the growth areas of the computing industry.

There is a need to study the way in which the technology can be exploited in government. Properly this would seem to be a role for the CCA, which at present does not have a relevant R & D capability. The CSERB should consider how it could best promote the use of computing in government, either by R & D under its own auspices, or by transferring a part of its R & D capability to the CCA.

## User support

It has been argued that the barriers to the use of the technology are not primarily technical. Nevertheless, there must still be a role for the provision of technical support and services to the user. This is a role that the NCC has been carrying out and there must be a strong case for extending the range of support to the NCC to enhance its current role, and to enable it to take the responsibility in a wider area than conventional data processing — specifically:

— the exploitation of the microcomputer by industry
— the use of word processing.

Care must be taken that in providing this support the NCC does not come into needless conflict with service industry. There would seem to be no case for the NCC to duplicate services like consultancy and software development which can be provided effectively by industry, since this weakens the development of an independent industry. There are, however, a large range of services of a more general and educational nature, which are not commercially viable, and yet are nonetheless economically important, and it is in these areas that the NCC should concentrate.

**Conclusions**

*The CSERB should undertake a radical review of its current support pattern, with the objectives:*

- *to switch support from the computer industry and computer-related R & D towards support of the use of computing*
- *to reduce the level of support within GRE's, releasing money and capability to promote activities outside the GRE's.*

*The ACTP and software products schemes should be reviewed with the intention of replacing them by a scheme, less oriented towards R & D and aimed at promoting specific uses of computing.*

*The function of the GREs should be reassessed and reoriented to provide support for the use of computing, specifically in the area:*

- *centre of expertise in semiconductor technology*
- *centre of expertise in languages*
- *monitoring of R & D in other countries*
- *R & D on standardisation*
- *policy forecasting*
- *use of computers in government*
- *general user support (NCC).*

## THE ORGANISATION OF R & D

The current arrangements for government R & D in computing are chaotic. Responsibility is spread between a number of organisations with no central policy, while within the ambit of the CSERB, activities are fragmented between a number of GRE's. If information technology is as significant as argued in this study, then there is a need for a central Institute of Information Technology to provide a focus for the subject.

### Industrial R & D

It is considered that an increased emphasis on the funding of industrial R & D would be inappropriate. In general, there is adequate technology available. What is required is the effective application of this technology. In this respect R & D support does not address the real problem, because it only funds the initial

217

1-10% of new product launching; also, it places emphasis on the technical aspects of the product rather than its commercial viability.

There is a need for support to companies who wish to introduce products which exploit information technology, in particular those which use microprocessors. This would seem to lie outside the role of the CSERB, both because the judgement to support such products should be commercial rather than technical, and because the sums of money involved would be beyond its budget. The CSERB should be urging the government to take appropriate action in this area.

### Military R & D

The CSERB already has a direct relationship with the Ministry of Defence (MoD) and is responsible for promoting commercial spin-off from defence work. It is considered that, in the future, there will be an increasing divergence between military and commercial requirements, and that the commercial exploitation of military work will become less relevant. The thrust of information technology is towards low-cost products and continuing usage. The direction of most military work is almost the converse, with the emphasis on performance and one-time operations. Thus, for example, the design of a commercial telecommunications network has totally different characteristics from those of a military network, which needs to be designed for a short, first-peak utilisation, and high unreliability (!) at the nodes.

### Academic R & D

Academic R & D is funded through the SRC. In the case of Computing Science, as with other engineering subjects, there is no direct relationship between the activities of the SRC Computing Science Committee and the CSERB. There is no formal liaison between the two funding bodies and no formal cross membership, and there are no assessors. At the informal level there is some interaction, particularly in the area of CSERB activities on Real-Time Technology and SRC activities on Distributed Computing Systems, but this has arisen by personal initiative. There is also informal interaction, in that the DoI is represented on both bodies, although by different individuals.

An examination of the activities supported by both bodies shows that there is no clear demarcation between the areas of responsibility. While the CSERB-supported activities may in general be more 'downstream' than those supported by the SRC, it is certainly the case that some of the research supported by the CSERB is more 'academic' than some of the SRC-supported research in the same areas.

The two main differences between the CSERB and the SRC lie not in the areas supported, but in the mechanistic aspects of support:

— The SRC acts as a sponsor, whereas the CSERB acts as a customer for research.

— The CSERB maintains research teams on a continuing basis through tenured posts. The function of the SRC is to supplement the sources of (academic) institutions on specific research topics for a limited duration, so that its research activities do not provide long-term security.

It is not clear that the CSERB funding approach is always the most appropriate for 'downstream' R & D, nor that the SRC approach is always the most appropriate for 'upstream' R & D. Indeed, as computing science (or information technology) matures, expertise becomes more specialised, and often large-scale investments in hardware, software, and knowhow must be made before useful results can accrue. This is making the SRC-sponsorship approach much less suitable in critical areas like communications and language.

There would seem to be a strong case for a much closer relationship between the funding activities of the CSERB and the relevant committees of the SRC (Computing Science, Electronics and Systems Engineering, Control Engineering). At the minimum, this might consist of an agreement on a common policy to be implemented by the two bodies within their own spheres of influence. More positively, it could be implemented by giving the CSERB control over some part of the funding dispensed by the relevant SRC committees (as happens with some other RR Boards), or it could be implemented by achieving some more satisfactory split of R & D activities — the SRC taking responsibility for 'upstream' research both in academic institutions and in GREs and the CSERB taking responsibility for 'downstream' research

219

both in GRE's and in academic institutions.

## Research associations
There is no research association for the computing industry. In the case of the hardware industry, this situation is considered reasonable, because it is dominated by one large company, while several of the other participants are subsidiaries or divisions of large companies. The position with the software industry is less satisfactory. The software industry is highly fragmented and there are few companies big enough to support ongoing R & D activity at any but the most pragmatic level.

The provision of funds for industry-oriented research in software (as opposed to product development, which is company specific and is already supported through the Software Products Scheme) needs more consideration. One possibility within the current scheme of things would be to give the software industry more direct control over the investment of money for software research at GRE's, either with or without some direct financial contribution from the software industry itself.

## Government research establishment R & D
On its inauguration, the CSERB inherited a mish mash of responsibilities, including the support of ongoing work in computing at the various GRE's. These perform miscellaneous R & D activities in a variety of areas with no apparent pattern of technological or geographic concentration.

The analysis on pp. 209 ff. indicates that much of the work supported within the GREs may not be of the greatest relevance. It is suggested that there needs to be a current review of current activities and a vigorous rationalisation in order to create a more effective approach for the future. As presently constituted, the GRE's appear to have no defined responsibilities, so that the CSERB must take the major role in determining the direction of funding. Given the effort available from a part-time voluntary group like the CSERB, this is not a satisfactory approach. A better method might be to identify specific mission areas for each GRE, and to make the individual GREs responsible for ensuring that their expenditures proposals provide satisfactory coverage of their mission areas. In this way, the role of the CSERB would be to identify the key areas for R & D, and then to monitor that the proposals in these areas were cost effective. This would seem to be more in line with the range of expertise and effort accessible to the CSERB.

**Institute for Information Technology**
The foregoing sections have discussed modifications to the support of R & D within the current overall pattern of support. There is, however, a case for a more radical approach.

While the U.K. has achieved some good results, at least in the R area of R & D, there are a number of obvious weaknesses which must lead to concern at the overall approach:
— lack of positive programme on R & D
— poor coordination between various research activities
— non-commercial relevance of some supported research
— poor technology transfer between research activities and industry.

Many of these problems may be blamed on the lack of a central focus for the subject, and could be resolved by the creation of an Institute for Information Technology. Such an Institute could act as a centre of technical expertise in the subject and might have the following responsibilities:
— provision of technological advice for government, particularly in the area of long-range forecasting for the subject
— promotion of the interchange information between research groups and industry (such an exchange would ideally be two way, and could be promoted by having short stay of part-time fellows from both industry and universities)
— active coordination of research groups in universities and elsewhere on the IRIA model
— provision of research association facilities to the software industry
— monitoring and dissemination of foreign development
— technical support for established standards
— research in key areas requiring ongoing research.

On the basis of this study, these last-mentioned key areas might include:

221

- standardisation activities
- languages
- industrial robotics
- semiconductor technology
- data communication.

Such research activities might include contract or sponsored research in universities and elsewhere.

Such an Institute could pull together the currently fragmented activities supported by the CSERB. It would be better still if it could have the active involvement of the SRC and the BPO, thereby creating a unified approach to the future of information technology.

## Conclusions

*The CSERB should re-evaluate the relevance of support for industrial R & D and should reconsider the extent to which the exploitation of military R & D is worthwhile.*

*The CSERB should consider whether it would be possible to develop a more coherent approach to the supportive research in conjunction with the relevant committees of the SRC.*

*The CSERB should consider whether it could establish more meaningful mission objectives for the GRE's, thereby delegating part of its responsibility for establishing a research programme and creating a more manageable situation.*

*The CSERB should consider the establishment of an Institute of Information Technology to provide a central focus for the subject.*

## THE ROLE OF THE NCC

This study has placed emphasis on the importance of the use of information technology as compared to its supply. In this context, the NCC is seen as having an increasingly important role to play.

### Support of the use of the technology

As programmable electronics and information technology become more pervasive, so it is the application of the technology rather than its provision that will have the greater economic importance to the U.K. Ultimately, we may expect the application to affect

222

virtually 100% of the economy, while the supply will affect only 1% or so. The prime emphasis should therefore be placed behind the application of the technology.

In this context the NCC is an important vehicle and, because of its unique nature, the U.K. has a positive advantage over the rest of the world. The CSERB should consider devoting a much larger proportion of its budget towards promoting the NCC and extending its activities in support of the user.

## Relation of the NCC and the proposed Institute of Information Technology

It is considered that the mission of the NCC and the mission of the proposed Institute are really quite different, and need to be sharply differentiated. The role of the NCC should be to support the user. The role of the Institute should be to maintain a technological capability to support the information industry. The distinction is best seen in an area like standardisation. Here, the NCC has an important role to play in representing the interests of the user; often it may find it necessary to help formulate the user's view in a technically complex area. The user's view however, is only one of many that must be taken into account in the standardisation process; thus it is undesirable that the NCC should be cast in the role of both judge and jury.

## Conclusions

*The CSERB should consider whether a greater proportion of its resources should be directed towards the NCC for the support of the use of information technology.*

# CHAPTER NINE

# POLICY IMPLICATIONS

Policy is about the future. It needs to be based on future opportunities, not on past failures. Policy formulation is, therefore, more difficult in times of rapid technological change such as is occurring now in the area of information technology.

The preceding chapters of this study have outlined a variety of developments that are anticipated as a consequence of technological change, and Chapter 8 has suggested how the CSERB might respond, in particular by creating an Institute of Information Technology, capable of assessing technological change and providing the basic inputs for policy making. This chapter reviews some of the broader aspects which it is believed may be impacted by information technology, and where, therefore, policy needs to be formulated.

## SCOPE OF POLICY
The main theme of this report has been that the current view of computing and computing policy is too limited. The developments of the technology mean that electronics, computing and communications need to be seen as interrelated aspects of a more

basic information technology. The reduction in cost of this technology means that its use will be pervasive, extending throughout commerce and industry, with far greater impact on the individual than has been the case in the past. For these reasons, policy will need to be formulated on a much wider basis.

It is unfortunate that direct financial support to one sector of the computer industry is often taken to constitute policy. Promotion of the computer industry is only one aspect of policy, and direct financial subvention is only one mechanism.

## The scope of the industry

The computing industry is only one element in the supply structure for information technology. The electronics industry, with its increasing focus on microelectronics as opposed to discrete components, will find that its products increasingly require computing techniques, even when not used in computers themselves. Similarly, the telecommunications industry will increasingly use computing techniques to implement its facilities, and will become increasingly involved in providing access to the services generated by computers. Policy needs to see these three subjects — electronics, computing and telecommunications — as aspects of a single information technology, which for the foreseeable future will be implemented by electronic techniques.

It must be expected that this convergence will have direct consequences on the structure of the supply industries involved, and that it will also cause industrial boundaries to be redrawn both upstream and downstream of the companies concerned.

Policy needs to be formulated across the three aspects of information technology. It is not a matter of coordinating the three but of ensuring that they are properly integrated and that the arbitrary barriers that exist, both in government and elsewhere, are eliminated.

## The need for a policy

The extent to which the government needs a policy towards information technology should be questioned. In that version of a free-market economy which exists in the U.K., the principle would appear to be that the government should interfere as little as is consistent with the achievement of its economic and social objectives.

Desirable or not, government involvement in the technology is already high and must be expected to increase. Because of the perceived strategic nature of the industry, it has been subject to government direction and financial support both in the U.K. and elsewhere. Similarly, the government — both directly and indirectly — represents one of the largest purchasers of the technology and, although its share may decline in the future, it would still be capable of asserting great influence over the industry by its procurement policy.

As yet there has been little government policy involvement in promoting the use of the technology or in controlling its consequences. The diffusion of the technology through industry and commerce must be expected to have considerable effect on the competitiveness of products and the efficiency of companies. As an exporting country, our economic position will be dependent increasingly on the extent to which we exploit the technology in competition with other industrialised countries. For this reason, it must be expected that government involvement will be forced to increase.

Finally, the consequences of the technology for the individual will have increasing importance. The exploitation of the technology may have a substantial effect on the pattern of employment and efficiency, leading to unemployment and inequality in the distribution of benefits. The scale of these effects will almost certainly warrant government recognition and action, and for this reason also government involvement in policy must be expected to increase.

## INDUSTRY VERSUS USE

Computer policy in the U.K., and in most of the other industrialised countries has been almost totally directed towards the promotion of the computer industry itself, which represents 0.3-1% of the total economy, or about 5% if relevant areas of electronics and telecommunications are included. Despite the growth of activity expected, this sector is unlikely to grow very much in economic terms because of the falling price of microelectronics.

By contrast, the contribution of computing and information technology to the rest of the economy must be expected to increase as the pervasive nature of microelectronics spreads the use of programmable electronics into wider areas of the economy. In the long term, it must be expected that almost every area of the economy will be affected by the use of microelectronics.

Under these conditions, it must be the exploitation of the technology which is the important factor to the U.K. The use of information technology will directly influence the efficiency of 95% of the economy, whereas support in the industry only affects 1-5% of the economy.

The present policy of support to the U.K. computing industry itself can act as a negative factor in promoting the use of the technology:

— By unduly restricting the freedom to purchase equipment, it may reduce the efficiency of critical parts of the economy.

— By attracting skilled people into the industry, it may reduce the number and quality of people available to use the technology.

While it is obvious that promotion of the use of technology must be the important aspect of future policy, such promotion presents a far more difficult problem simply because of the pervasive nature of the applications. Three courses of action appear open.

— promotion of specific applications of the technology
— financial inducements to use the technology
— education.

At present there would appear to be a considerable risk that the U.K. will fail to exploit the technology as aggressively as our industrial competitors. The consequences of this could be extremely serious since the U.K. largely depends on the export of industrial products, and could become uncompetitive if it does not take advantage of the technology. It would seem, therefore, that all three methods of promotion should be considered as part of a policy on information technology.

So far as promotion of specific applications is concerned, these are only of interest where they relate to products — either software and knowhow products  created by the computing industry for sale to the rest of the economy, or products which embody the technology, like improved consumer durables. Thus,

227

such promotion divides into two parts, support of the software and system industry to sell capability to industry, and support of specific industrial products that use programmable electronic technology.

## EFFECTIVE LEVELS OF SUPPORT FOR THE SUPPLY INDUSTRY

One of the most important effects of policy is that it draws attention. Merely by selecting a specific area for action, policy focuses attention on that area, and the resultant momentum derived from a small shift in direction by many companies and organisations can far outweigh the specific action or support provided by government.

Nevertheless, in order for government investment to be effective, it must be made at a credible level. The computing industry has to compete on an international basis, and in judging credible levels for support, these must be compared against the levels of investment being made by other countries, like Japan, or the FRG. A selective promotion of this kind needs to be spelt out clearly, so that the computing industry can make the appropriate management judgments.

At present, the support programme of the government is heavily biased towards R & D, but it is not clear that an adequate proportion of this work is translated into commercially viable products. One possible reason for this emphasis on R & D is that the early stages of a product cost much less to support, so that this approach is a method of apparently making a limited level of support stretch further. But support in the early phases is wasted, unless the money is available to support the marketing phase. It would appear that there is a strong argument for shifting the balance of government support from the early phases of a project to the later phases:

— Selection of worthwhile projects at the early stages is far more difficult, and is an inappropriate task for government.
— Monitoring and control of investment at the R & D stage is far more difficult for government than monitoring of the marketing phase.

— If a company needs support for the early phase, it may
    not be able to make the product without further support.
The implications of this are again that given the current level
of support to the computing industry, it needs to be targeted
more specifically on a narrow range of activities, and again this
implies a clearly defined and publicised policy.

The question whether the total level of support should be
increased or not is a far more difficult matter. This needs to
be considered not in isolation, but in comparison with alternative
investments that can be made by government. The primary
importance of computing is seen to be in its use rather than in its
supply, so that decisions to support the supply industry should be
based on a view that this will, in itself, provide a better return
than the support of other areas of the economy, for example
agriculture.

If the U.K. does wish to compete effectively in the information
technology market, the level of support provided to its industry
will have to be comparable with that provided in the other major
countries, and this implies increasing the level of investment
by an order of magnitude. The alternatives are either to abandon
investment in the area or to adopt a coherent policy of elective
support.

## COMMUNICATIONS

It becomes clear from this study that communications is the
underlying technique necessary to make most aspects of information technology feasible. It is also apparent that the current
rate of progress towards a suitable communication system by the
BPO and the PTTs elsewhere is inadequate, and that the resultant
lack of suitable communications services is likely to be the main
constraint on the development of information technology. It
follows that the activities of the BPO need to be fully integrated
with the development of information technology, and this means
that policies towards the technology must include policy towards
the BPO.

The provision of appropriate telecommunication facilities
could be the most effective way to promote information technology in the U.K. An active investment programme in this

area has a number of advantages over and above the basic capability that it creates:

— Any benefits accrue directly to the U.K., giving a clear competitive advantage.
— The BPO is already the single largest user of the technology. A programme of positive investment in advanced technology by the BPO would provide a considerable stimulus to the information technology industry as a whole.
— By creating a market for advanced telecommunication products, such an investment programme would build a market leadership position in this area for the U.K.

The BPO is seen as the major employer directly affected by the introduction of information technology. Improved technology can lead to a direct reduction of labour in the area of telecommunications operations due to higher reliability and simplification of equipment, while it also reduces the labour content in telecommunications manufacture. The use of electronic mail would cause a switch from postal services to telecommunication and would also have a considerable effect on employment levels.

For all these reasons, the attitude of the BPO is seen as the key to the future of information technology in the U.K., and the BPO must, therefore, become an integral part of any policy towards information technology.


## EDUCATION

A policy towards information technology must also encompass at least these aspects of education:

— vocational training to exploit information technology
— adult re-education programmes to cope with changing labour patterns as a consequence of the technology
— contextual education to ensure that everyone is aware of of the technology and its potential consequences.

In the short term, there would seem to be a serious training problem relating to the exploitation of microcomputers. At present, the microprocessor requires a combination of application, computing and electronic skills which, in general, have not been taught together in universities. To ensure that the microcomputer is properly exploited, it will be necessary to provide engineers with the relevant skills and to educate management to appreciate the potential of the microcomputer in diverse industries. To meet these various needs, a programme is required to:

— generate relevant degree courses in universities
— provide retraining for engineers and programmers in industry
— orient management and unions towards the opportunities the new technology presents

Such a programme must be regarded as urgent.

The development of information technology is changing the requirements for education. It is little use providing narrow courses in computing or in electronics or telecommunications. What is required is the synthesis of these subjects to provide degree courses of greater relevance to the future. One must also question the extent to which the current specialist education really has much relevance to the future. The number of engineers who will participate directly in the development of the technology is small in comparison with the very much greater number who will be involved in its exploitation. For the present, the use of a device like a microprocessor requires considerable specialist skill, but it must be expected that, as its technology matures, the degree of specialist skill required will dwindle to a negligible level. Similar remarks can be made about programming itself. At present this requires specialist techniques which in due course will disappear. What is required is not so much specialist teaching in the area of computing and information technology, as the teaching of the concepts of computing and programming techniques to a much greater proportion of university undergraduates. It should perhaps be a target that every graduate has the capability to use computer systems and a thorough understanding of their potential.

In due course, information technology is likely to restructure the jobs of many people, either by changing the way that the job is done, or even by eliminating it altogether. This situation will place a great deal of emphasis on adult education and retraining, which will become far more important in the future. It will be necessary to provide greatly improved facilities for adult education and to make the concept more socially acceptable.

Finally, in the long term, information technology could have a substantial impact on the basics of teaching. Arithmetic will be changed both because mechanical aspects of computation are eliminated and because the underlying concepts of programming are so basic to the thinking process that they should be introduced at an early age. Writing in its current form will be eliminated in favour of direct keyboard entry and editing, while reading will be enhanced, because the choice and access to information takes on an increasingly important role. Perhaps most exciting of all is the possibility that the information systems themselves will be able to play a far more positive part in the teaching process.

## SOCIAL POLICY

The implications of much of this study are that the working through of the technology will have major social consequences. In Chapter 3, the Information Revolution was compared with the Industrial Revolution; while the level of industrialisation is now much higher, so that the impact my be correspondingly reduced, there can be little doubt that the technology will generate considerable social change.

As was the case with the Industrial Revolution, the improvement in productivity that information technology will allow is inherently 'good', because it can be converted into greater leisure or a higher standard of living. However, unless it is properly planned, the short-term consequences could be disastrous, because the demand side of the economy has a much slower response than the supply side, so that rapidly increasing productivity leads, at least initially, to higher unemployment and the unequal distribution of wealth. The concern must be that unless these changes are foreseen and adequately controlled, there may be serious untoward consequences and the long-term benefits of the technology may be lost.

The two major consequences which need to be taken account of by social policies are:
  — The increased level of unemployment that could result from the adoption of the technology. The objective of policy should not be to prevent labour displacement,

but to make it acceptable, by ensuring that it does not convey social hardship or stigma, and by providing the people involved with creative opportunities for the future.

— Inequality of earnings. If the productivity of a particular job is greatly increased in comparison with other jobs, it can be done with less effort or time. The consequence is that some jobs will be paid disproportionately, for the effort or skill involved, in comparison with others which have been unable to benefit from the technology. The same situation recurs on the macro scale with companies. Some sectors of the economy will be able to use the technology to make disproportionate profits, while other sectors will be unable to benefit to any significant extent. In the long run, the operation of the free-market economy removes such anomolies. However, in the short run, positive government policies towards individuals and towards companies may be necessary if some are not to be penalised unduly.

The technology will also offer changes to the mode of government. More effective information collection and analysis can improve the efficiency of government and perhaps simplify the interactions between government and the individual. While it is necessary to consider the adverse effects of such developments — for example in the area of privacy — equal emphasis should also be placed on the positive benefits to be obtained, so that a balanced judgement can be made. Particular aspects where information technology might impinge on government and where policy decisions will be required are:

— participation by the individual in decision making, by improved systems of voting, consultation, etc.
— simplification of record collection and keeping
— provision of different (and, ideally, simpler and more cost effective) methods of taxation and tax collection
— changes in the law relating to information, much of which is based on the limitations imposed by existing paper systems.

## POLICY IMPLEMENTATION

There are a large number of institutions in the U.K. concerned with various aspects of information technology. These include, in the public sector:

- Computers, Systems and Electronics Requirements Board of the Department of Industry
- Computers, Systems and Electronics division of the Department of Industry
- Various National Economic Development Office groups
- British Post Office
- National Computing Centre
- Central Computing Agency
- Science Research Council
- University Computing Boards.

and in professional or trade groupings:

- British Computer Society
- Computing Services Association
- British Electrical and Allied Manufacturers Association.

Associated with the various groupings of these institutions are innumerable committees concerned with coordinating various activities. What is totally lacking is any institution with overall responsibility for creating and implementing the policy towards information technology, covering both the supply and the use of the technology. Until such an institution is created with the necessary authority to integrate, rather than merely coordinate, the diverse activities currently engaged in, it is unlikely that an effective policy will emerge.

In the future, information will be as important a resource as energy, and its proper provision, control and exploitation will be a major factor in the economic well-being of the U.K. Management of this resource is important enough to demand consideration at the highest levels of government. What is required is the first step towards the creation of an integrated policy, with the appointment of a Minister within the Department of Industry with specific responsibility for information technology.

# GLOSSARY

**Access Time**.     Time taken to call for data from a storage device and place it where required.

**Active Storage**.     Memory requiring power in order to retain its contents.

**Address Space**.     The number of separate locations provided or occupied by a device.

**Alphanumeric**.     The class of characters including numbers and alphabetic characters and certain punctuation marks.

**Architecture**.     The structure of the relationship of the various facilities and sub-facilities in the system.

**Archival Storage**.     Extended storage such as that used for very large files.

**Balanced Pair Lines**.     A pair of parallel wires carrying differential signals and useful where rejection of interference is important.

**Bandwidth**.    The frequency range available from or used by a system, e.g. image transmission such as T.V. requires a higher bandwidth than a voice or data channel.

**Baud**.    Essentially the same as BPS.

**Binary Representation**.    Representation of characters or a set of characters, e.g. alphanumeric or other symbols, in bit form.

**Binding**.    The act of converting the symbolic address in the same language to the address used in the running program.

**Bit**.    An abbreviation of binary digit, one of the two values (0 and 1) used in binary notation. The term is extended to include its representation, e.g. a magnetised or non-magnetised spot on a recording surface.

**bps**.    Bits per second, a unit used to measure the speed of transmission in a telecommunication channel.

**Bubble Memory**.    Type of memory in which data is stored on moving bubbles of magnetism on a semiconductor, this storage being nonvolatile.

**Bus**.    A passive interconnection system where the devices are connected in parallel, sharing each wire with all other devices on the 'bus'.

**Byte**.    An eight bit binary number, which is used to represent a symbol capable of being coded within its capacity, i.e. a range of $2^8$ symbols.

**Cache Storage**.    A relatively high speed store intermediate between a central processor and its main memory.

**CCD**.    Charge-coupled device used as the basis for switching circuits.

**Ceefax.** The BBC's name for their consumer information service involving the calling up of a BBC TV screen of preset broadcast pages of information. Generic term: 'Teletext'.

**Clock Generation.** The generation of electronic pulses repetitively in order to synchronise all other operations.

**Coaxial.** On the same axis, usually referring to parallel cabling.

**Compiler.** A program which converts computer instructions written in a *source* language into *machine* code.

**Compiler Time.** Time taken by the compiler to convert the source language into machine code.

**CRT.** Cathode ray tube.

**Cryogenics.** The use of extreme cold, often near absolute zero, i.e. − 273°C, in order to make use of the phenomena of superconductivity, particularly advantageous for very fast micro-electronic circuits.

**Daisy Wheel Printer.** The type head of a daisy wheel printer is a circular arrangement of all the characters stamped onto the end of stalks. When a character is to be printed, the head spins and the correct character is pressed down onto the ribbon.

**Decimal Representation.** Using characters in the range 0-9 to represent symbols.

**Dedicated Ports.** An access point, to a communication channel, used only for one specified type of traffic.

**Digital Communication.** Sending messages where successive characters are represented by discreet and separate values of the representation. Commonly refers to telecommunications links.

**Discretionary Addressing.** The avoidance of faulty areas of a storage medium by ensuring that the area of storage addressed is other than that associated with the addresses of faulty areas.

**Dissipation.** Power consumed or lost along a channel.

**Do Loop.** An iterative process in a computer programme usually specified by the nature of the iterative computation and the extent of the iteration repetition.

**Drive Circuitry.** Any circuit generating pulses for operating some electro-magnetic device.

**Electron Beam Machining.** A method of achieving a lineal pattern on a material by directing a focussed electron beam on to the surface.

**EMS.** Electro-magnetic spectrum.

**Facsimile Transcription.** A method for transmission of visual data.

**Floating Point Arithmetic.** A calculating method where numbers are held in mantissa + exponent form, e.g. using 17E6 for 17000000.

**Floppy Discs.** Flexible ferrite discs used for mass storage in computer systems.

**Frequency Multiplexing.** The division of the electro-magnetic spectrum into bands, and the allocation of those bands to functions. Since radio stations occupy different bands they may be 'demultiplexed' by tuning into them.

**Front End.** The input section of a system.

**High Level Language.** A language allowing the programmer to write programmes in powerful 'English type' statements or commands.

**Holographic Store.**    A method of data storage using laser technology.

**Ink Jet Technology.**    A method of printing where an ink jet is used to draw characters instead of the more usual impact or thermal methods.

**Instruction Set.**    The list of all possible instructions that a computer may execute at the machine level (the lowest level).

**Integrated Circuit.**    A semiconductor chip holding a circuit formed of transistors, resistors, diodes etc.

**Interface.**    1. An imaginary line drawn between two portions of a system in order to highlight differences between them. 2. A circuit or program designed to handle or take account of discontinuities in a system.

**Internal Refresh.**    Self contained system for maintaining data in a dynamic storage medium. i.e. a storage medium which would otherwise lose its pattern without replenishment.

**Interrupts.**    A method of asynchronously demanding the attention of a computer.

**Ion Implantation.**    A method of doping semiconductors or metals by bombardment with fast moving IONS.

**Josephson Device.**    A semiconductor device operated at cryogenic temperatures, making use of the physical properties of individual atoms and molecules instead of the bulk properties of the materials used. Such devices offer extreme packing densities and switching speeds well beyond current capability.

**K-Storage.**    Shorthand method of representing the number 1024 ($2^{10}$) in relation to extent of storage. e.g. 4k = 4096. Commonly used as shorthand for thousands (000s).

**LCD.** A Liquid Crystal Display. Works on the principle that the liquid crystal medium can be made to polarize light by placing an electrical potential across it.

**Magnetic Media.** A medium which stereo data using the principles of electro-magnetism.

**Matrix Printer.** A printer in which each character takes the form of a pattern of dots produced by a stylus or number of styli moving over the surface of the paper.

**Micro-Computer.** A single silicon chip containing a micro-processor as well as memory and perhaps some input/output facilities.

**Micro-Processor.** The central processing unit of a computer implemented on a single silicon chip.

**Mini-Computer.** Originally a computer significantly smaller in size, capacity and software capability than the larger main-frame computers with which it is contrasted. The differences have been blurred by technical advances in time.

**MOS.** Metal oxide semiconductor field effect transistor. Very low power devices which facilitate very high packing densities in digital integrated circuits.

**Multiplexing.** Pertaining to any system in which a single device is used for many purposes, but usually referring to communication channels.

**N MOS** (cf P MOS). N channel metal oxide semiconductor field effect transistor as used in both analog and digital circuits. P & N refer to desirable but distinct solid state effects sort.

**Operands.** The item in an operation from which the result is obtained by means of defined actions. Programming manipulates data via its operand.

**PABX.** Private automatic branch telephone exchange.

**Parallelism.** Simultaneous, not necessarily co-ordinated but still controlled, execution of data manipulation.

**Photolithography.** A method of achieving a two-dimensional pattern on a suitable surface by using either an optical mask or direct copying from a two-dimensional pattern.

**Pinout.** The specification of the function and position of each pin or lead of an integrated circuit.

**Plug Compatible.** Two devices which will fit into the same socket are said to be 'plug compatible'.

**P MOS** (cf N MOS). P Channel metal oxide semiconductor effect transistor as used in both analog and digital circuits.

**Printer.** An output device which converts data into a printed form.

**Programming Language.** The set of instructions understood by the computer system.

**RAM.** Generally accepts to mean 'Random Access Memories', in which any location may be written to or read from in a Random Access fashion.

**Random Access Storage.** A storage device to give a constant access time for any location addressed independent of the last location addressed. This should be contrasted with serial access where items are located sequentially.

**Raster Scan.** The method used in Cathode Ray tubes to place an image on the screen.

**Resolution.** The smallest possible increment of a valve.

**Ring Main**. A system where a power supply wire is connected at both ends to the power supply, allowing a localised load of twice the rating of the wire.

**ROM**. A read only memory contains permanent data that can be read but not altered.

**Run Time**. The time during or at which a program is run.

**Semiconductor**. Any material which behaves in the following way. It acts as a conductor of electrical current when the voltage across it is above a certain level. It acts as a resistor when the voltage is below that level.

**Serial Transmission**. Transmission of data in which each unit of data being transferred travels in sequence.

**Space Domain Multiplex**. Physical allocation of physical resources to different physical activities for a period of time. cf. time domain multiplex.

**Sparse Matrices**. A sparse matrix is one in which one or more of the elements are undefined or zero.

**Time Domain Multiplex**. Sharing of a single physical resource by the allocation in an interleaved manner of physical activities each occupying inter-leaved time slots.

**Transducer**. A device which converts energy from one form to another.

**TTL Logic**. Transistor Transister logic. A particular method of fabricating logic circuits.

**Virtual Storage**. A store management system enables a user to use the storage resources of a computer without regard to constraints imposed by a limited main store.

**Viewdata**. Currently called Prestel. A G.P.O. system which enables a T.V. user to communicate via a telephone line to a main computer system.

**Virtual Processor.**   A fictitious machine created by software in order to simplify programming.

**VMOS.**   A technique which increases the packing density of metal aside semi-conductor field effect transistors by placing them in a V groove cut in the device substrata.

**Wafer.**   A slice of semiconductor material used in the manufacture of integrated circuits.

**Wiegand Effect.**   The propagation of a square pulse along a doped alloy wire associated with a changing magnetic flux at one end.

**Williams Tube.**   An electrostatic storage cathode ray tube.

**Word.**   A basic unit of data in a computer memory — consists of a predetermined number of binary digits.

**Xerographic Printer.**   A page printer in which the data is electrostatically placed on a plate. The plate is covered with a resinous powder — which adheres to the uncharged regions only. The resin is then transferred to another medium such as paper.